# Nature Rambles

Norman L. Davis

INFINITY
PUBLISHING

ISBN 0-7414-5959-0

Printed in the United States of America

Published December 2010

INFINITY PUBLISHING
1094 New DeHaven Street, Suite 100
West Conshohocken, PA 19428-2713
Toll-free (877) BUY BOOK
Local Phone (610) 941-9999
Fax (610) 941-9959
Info@buybooksontheweb.com
www.buybooksontheweb.com

#### Acknowledgements:

Typing, editing and formatting help from the staff at From My Shelf bookstore, in Wellsboro, PA, including:

*Kasey Cox*

*Kevin Coolidge*

*Jen Colson*

*James Yskamp*

*Kristin Stam*

As well as community members:

*Gabe Hakvaag*

*Donna LeSchander*

*Bryan Robinson*

Photo credits:

Front cover – *Norman Davis*

Back cover – *Janice Davis*

Artwork by the author

To the Memory of My Mother and My Father,

Viola and Maynard Davis

Who early in life led me on many a ramble afield

and introduced me to the world of Nature

*Of 186 original essays, there are 155 in this volume.*
*Altogether, they span a period of three years.*

# Foreword

Today we are living in a high-tech society where our lives are full of activity and stress. We live in an unnatural environment we have created for ourselves, and it is taking its toll.

We need to take time to leave the artificial behind and find serenity for the soul. This can only be found in the natural environment we were designed for.

The essays in this book are intended to do just that, especially for those whose circumstances preclude actually roaming afield. Each one will take the reader into the countryside and introduce him or her to the wonders – yes, even miracles – of the natural world. To many, it will be like opening a door to a realm never before explored and their horizons will expand as they listen and watch with the writer, as together, they follow him afield.

These essays start and end with the glorious days of Autumn, with all the wonders of the delightful seasons sandwiched in between.

This said, let's head out in our first ramble!

--N.L.D.

# AUTUMN

Once again the length of day and night are equal. Summer is over and autumn has begun. Slowly and ever so surely the warmth of summer is retreating to be replaced by cold nights and cool days although a countryman looks forward to many mellow, warm days yet to come.

In late winter, when the vernal equinox heralded the arrival of Spring, a countryman watched as the lengthening days softened the raw, sharp edges of Winter and the wild creatures became more active as they too, looked forward to the time of their own renewal.

Now the pendulum of the seasons swings ever farther in the opposite direction. Canadian geese set their wings and honk resonantly as they prepare to alight on the nearby lake. The smaller birds that spent their summer and nested in the north-country are moving southward a few miles every day now.

On an overcast morning with only a hint of a breeze ruffling the leaves, a countryman observed several species migrating or preparing to do so. Among these was a Nashville warbler, a wood pewee and a black-throated green warbler flitting in and out of the branches of a willow tree after their insect prey. The pewee used out-hanging branches as observation perches from which it sallied forth every now and then as an insect came its way.

Several tufted titmice foraged the woods-edge and when they reached a clump of sumacs they paused to sample some of its tart red fruit, conversing companionably back and forth. A chickadee cousin called out cheerfully to them as they passed by its perch in a wild apple tree.

A family of flickers in an ash tree call back and forth to each other -- "wick-up, wick-up, wick-up," as if seeing who could be the most polite. A red-bellied woodpecker hitches his way up a utility pole, no doubt wondering what happened to the limbs. Finding no food, it swoops down to a fencepost and with no better luck there flies into a woodland where its hoarse "chaw, chaw" call can be heard.

The red-bellied woodpecker only has a touch of rose along its belly, its crown and nape, on the male, is crimson, while the female has red at her nape only. These zebra-striped woodpeckers are one of several species that have worked their way northward from the southland over the past years, no doubt aided by the proliferation of bird feeders.

Two killdeers cry mournfully overhead as a crow conference convenes in the field below. These farmland plovers will soon be heading south to spend the winter in a warmer clime.

The leaves of the smaller trees are exchanging their shades of green for their colorful autumn hues, while the fall flowers bloom in profusion across the fields and along the roadsides as if in realization that their time in the sun is of short duration.

Clumps of yellow-petaled thin-leafed cone-flowers brighten otherwise drab areas of field and wayside.

Several bluebirds adorn the tops of pastureland fenceposts as they scan the pasture for insects. Their calls, carried by a busy breeze, do not have the cheerful note of anticipation that they had in March. Rather, their calls seem to be a somber, sad farewell to summer.

The early frosts of late September have left their mark upon the landscape. As September bows to October the once-proud flowering plants one by one fade away and there is little left but dark stems and stalks, their once green and flourishing leaves now droop, lifeless, wrinkled and brown. Where bees once hummed among bright blooms, now finches, sparrows and other seed eaters feast on fresh, new seeds.

Goldenrod is now tinged with greenish-brown, matching the changing suits of the goldfinches. The sumacs' flaming foliage and the carmine colors of the Virginia creeper hide the cardinals as they feed on their fresh fruit.

Catbirds still mew from the thickets, reluctant to let summer slip away. A phoebe, also reluctant to leave, calls from its perch on an exposed branch. Its tail alternately lifts and drops, a characteristic that even at a distance distinguishes it from its relative, the pewee. In a few more days it will no doubt take its leave where food will be more abundant. A countryman knows that when once more he hears its lively call that Spring has arrived again. The phoebe, the bluebird and the tree swallow follow closely behind the robin to announce the vernal season.

Ripening grapes fill the air with their fragrance, as vines hang heavy with their clusters. The sun and the showers have completed their tasks of bringing them to maturity. All too soon, the freezing fingers of frost will be plucking their foliage, exposing the unharvested fruit to frost's next onslaught.

Where once apple blossoms filled the air with their fragrance, there now hang spheres of varying hues of red. They also are ready for the harvesters. Those that have fallen to the ground provide a succulent feast for deer and jays as well as other feathered and furred creatures.

Yellow jackets busily extract sweet juices from broken and bruised areas of the fallen fruit. A countryman knows that when he sorts through them for a tasty pie that he must at least show an element of respect for the short-tempered wasps.

Wild grapes hang in loose bunches while Royal blue berries contrast sharply with the sanguine leaves of the Virginia creeper. These and many other small fruits provide ample fare for lingering songbirds.

Later, after the frosts and wind and sun have had their way with them, their withered and perhaps dried remains will provide yet more sustenance for those hardy feathered and furred folk during the long, harsh winter.

October is the summation of the seasons, a time of fruition and harvest and a time of storing away Nature's bounties. It is the season of the year when a countryman senses an even closer tie with the land, for this is the culmination of all the growth and fruition that has gone before.

We need to attune ourselves to the rhythm of the seasons and listen to the many voices of Nature. It will prove a soothing balm to the soul, and one will find relief from the tensions and stress of the world around us.

A calmness and a sense of serenity seems to pervade the countryside. It is a time of fullness, a time of completion. A hush hangs over the land as if in anticipation. It seems as though all creation is relaxing and taking a bit of a vacation before settling down to the hustle and bustle of preparing for the cold stormy months ahead.

A few bees and wasps busy themselves among the late-flowering plants as do other insects.

For the most part, the birds are quiet. Most of the summer residents have departed and their voices no longer are heard. Those who remain give only call notes or a brief version of their songs. Even the blue jays and crows are more quiet, although their raucous calls interrupt the tranquil environment now and then. If they can find a hawk or owl to pester then their persistent voices can be heard for an hour or more. Squirrels busy themselves cutting nuts from nut tree branches.

The squirrels and jays appear to each other as rivals, although usually there is more than enough acorns and nuts and field corn to go around. They don't think so. On occasion, the woods will be filled with their noisy exclamations; a squirrel proclaiming its rights of possession, emphasized by a violently jerking tail, while one or more jays will denounce the squirrel as a bushy-tailed thief. Then, as if forgetting what all the fuss was about, both parties go about their business and the woodlot is once more quiet.

Two dragonflies swoop and hover in the bright sunshine over a field of tarnished goldenrod and purple asters. Soon their lives will be over, but this day they are enjoying the warmth of the sun. We too should appreciate and enjoy each day as it is given to us.

A field sparrow feeds quietly with several song sparrows amongst the weeds. Its voice is no longer heard ringing out over vineyard and field while cicadas buzzed from their leafy domain. Now both are silent. The sparrow is still here, but the insect, having accomplished its purpose in life, has long since left the scene of action.

White-throated sparrows have drifted down from their northern summer homes. They call, tentatively, from the bushes and under-brush, their voices muted. The weather will be milder here and their food more available throughout the winter.

A female yellowthroat peeks out at a countryman as he watches an immature male yellow-bellied sapsucker excavate its small, oblong holes high up in a large willow. These holes, made in living wood, fill with sap and lure insects which, in turn, provide sustenance for the woodpecker. Many hundreds of holes are made in many different kinds of trees so that there are many sources of food. The woodpecker does not suck the sap as its name implies, but on occasion can be injurious to fruit trees. However, they are not a common bird.

Hummingbirds often frequent the sapsucker's diggings as a source of energy and protein, sipping sap and dining on small insects.

Small flocks of geese now can be seen; some may be resident birds, but others fly high and purposefully southward, honking their farewell in passing. A countryman knows that the larger flocks will not be too far behind, but for now it is a time of enjoyment.

A countryman thinks Autumn is one of the finest seasons of the year. Each day can be full of delightful surprises no matter how the day may commence.

One such day starts drab and gray, low clouds veil the face of the early morning sun while thin clouds scud across a full moon that hangs in the western sky.

One after another, small flocks of geese appear in the northern sky, their voices preceding their actual presence. They pass overhead, fore-runners of events to come.

A crow sentinel sits atop a distant tree and surveys a field of goldenrod, its now-muted colors brightened here and there with clumps of deep purple asters. A few bright blue blossoms of chicory remain to add additional color to the scene.

At one edge of the field, white asters add a frothy border while across the field a doe and her fawn feed on clumps of grass growing among the weeds. The fawn has long since lost its spots and forages beside its mother. It has grown to be two-thirds the size of its dam. Each deer is wearing its winter coat of grayish-brown that blends well into the field in which they feed.

The summer robins have gone, but have been replaced, to a lesser extent, by those that nested in the north country. Many of them will winter in the area, subsisting on fruits and berries in the orchards, vineyards and hedgerows. This day they chirp and carol pleasantly from their tree-top perches. A goldfinch softly calls "per-chic-o-ree" as it dips and glides along its way.

A subdued atmosphere pervades the landscape. Sounds are muted and the usually bright foliage is transformed into pastel shades. There is an earthy odor of damp weed and flower stalks, of leaves and moist soil. A short, sudden breeze carries the heady fragrance of fully ripened grapes.

As the day progresses, the sky clears and only scattered white clouds form against the azure sky. The sun warms the land. A woodchuck eats voraciously, adding a little more fat to its already rotund form. In a few more weeks it will retire to its burrow and commence its long winter sleep.

The insect chorus, although diminished in volume, plays a gentle background to crickets who lively chirp in the afternoon warmth.

Shadows lengthen across the land and the sun slowly sinks out of sight. Darkness settles quickly, like a soft, comforting blanket, over woodlot and field. A countryman enjoys this portion of the day fully as much as the earlier one.

He stands on a knoll with a commanding view of the surrounding countryside. There are no artificial light sources to distract or distort his view.

The darkness deepens, and overhead myriads of stars shine brightly, tracing outlines of numerous constellations. The Great Bear, or Big Dipper, lies to the north and dutifully points to the North Star.

A goose voice drifts earthward in the mellow light of the full moon which bathes the landscape. A katydid rasps its name in the darkness; crickets chirp slowly now as the temperature drops, the cadence of their calls seems to echo the heartbeat of the earth.

Another autumn day surrenders to the wonders of the night.

A countryman enjoys October, he thinks, as much as he does the month of June, yet in a vastly different way. June is a time of freshness, new beginnings and new life. The autumnal month is a time of finishing, a time of ripeness and maturity, a season of fulfillment.

The hickories, walnuts, butternuts, beeches and oaks all beckon to be harvested, secure in the knowledge that sufficient fruits will remain or be buried and forgotten so as to perpetuate the species in future generations of seedlings.

As the years progress, some of these seedlings will survive to become saplings and of these, a few will eventually become trees which will, in turn, reach maturity and produce nuts and acorns to repeat the cycle once more.

The pines, spruces and firs are laden with their cones also. Numerous birds and red squirrels relish the seeds imbedded within the cones. Some birds, like the crossbills, depend on those seeds for their sustenance.

Now the nut trees stand starkly outlined against the deep azure of the sky, or perhaps silhouetted against the dark greenery of the conifers.

A countryman revels in the bright blue skies and the mellow warmth of the days that seem to whisper that summer has returned for a brief visit. It is usually after the first frosts that days like these arrive to bring pleasure to the soul. Then there are the days when the sun filters down through a lightly overcast sky and a gentle south wind brings added warmth to the countryside.

Bumblebees busy themselves among the white and purple of the asters, while yellow sulphur butterflies flit and flutter in the warm, caressing breeze. In the background, the somnolent voices of remaining summer insects complete the autumn scene.

Windy and showery days have done their work among the brightly colored trees, many of which have been left with perhaps only the lower portion with leaves, like a many-colored skirt still resplendent in the autumn sun. The ground beneath them is a veritable palette of oranges, reds and yellows. Then there are other trees that were not ready to release their leaves during the stormy hours and now they appear, full bodied, in all their regal splendor.

Milkweed pods have split open, spilling their contents to any willing breeze. Each diminutive seed is attached to a silken parachute that may carry it several miles from its maternal pouch. Already, many have started their journey and cling to weed stalks several feet away. During the wind-filled months ahead they will be dissipated across the surrounding landscape. A few will take root in the spring; others will provide food for birds while the fluffy floss will serve small mammals as soft, heat retaining bedding through the long, cold winter.

In all these observations, a countryman can see the hand of a loving, caring Creator that sustains all life in its varied forms. The natural world is not left to its own devices but is maintained and sustained by the One who designed it. We, also, are a part of this natural world and as such are assured of His constant care.

The autumn rains bring a measure of relief to the land. They help restore the water levels to farm ponds and give an added impetus to the springs that did not dry up during the long, hot summer. Small streamlets have re-appeared and provide the wild folk with water.

Along with the rain, the winds have stripped bare many of the trees while those that escaped the onslaught stand proudly in their colorful autumn robes. A few trees seem reluctant to lose their summer green, but they will as the season progresses. Even now, there is a yellowish cast to the foliage as the chlorophyll slowly retreats from the leaves.

Ox-eye daisies bloom in low, scattered clumps at the edge of a field while a few bright yellow dandelions keep them company. Sheltered by the nearby hedgerow and warmed by the afternoon sun, they continue to cheer the passer-by while their less favored companions have long since gone to seed.

The blossoms of the purple asters have worked their way up their stems and soon only their seed containers will remain. Here and there a tall stalk of evening primrose can be seen with bright yellow blooms at the very top while beneath them are seed pods all dried and brown. From a distance the plant appears like a huge candlestick with the yellow flame burning at the top.

There are gray days with chilling winds now, but then a day will dawn clear and the sky turns a dark blue, and a gentle southwest breeze flows over the land. A few errant warblers feed in the tree tops as they journey southward. The warming day elicits some soft, sweet warbles from a bluebird, but querulous cries of the catbird no longer are heard from the shrubs and bushes. It has finally departed to a more congenial locale.

A countryman enjoys walking through the newly fallen leaves and listening to their rustling, crunching sounds. This enjoyment is further enhanced by earthy, somewhat spicy odors released by the crushed leaves. We also should learn how to let fragrance flow from our lives even though we may be crushed by the vagaries of life.

The sounds and scents of the leaves now remind a countryman of days of yore when, as a lad, he trudged along a winding path through a woodlot to the country school. The sounds and odors today bring to mind memories of a kinder, gentler time, a time that was slower-paced and so much less complicated, a time that allowed for periods of quiet contemplation.

As these memories flash across the screen of the mind, he once again hears the incessant chucking of the chipmunks. He hears red squirrels scolding in the hemlocks. He also hears a partridge scuffing along its way in the fragrant leaves until, startled by his appearance, it roars into flight, zig-zagging among the tree trunks and out of sight.

The voice of a cock pheasant rings out and interrupts a man's reverie and his thoughts return to the present. As he rambles along, two hen pheasants rise steeply skyward and disappear on whirring wings.

A countryman knows days like this cannot last, but he also knows that each succeeding day holds some treasure to be sought out.

Red-breasted nuthatches forage among long, brown spruce cones, their voices sounding like small tin horns. They are amiable little chaps and their company is always enjoyed.

The afternoon sun warms the land and one's body and soul. Myriads of tiny insects dance frenetically in the air. Their time in the sun will all too soon draw to a close, their purpose in life fulfilled.

There is a seasonal "changing of the guard" in late October and early November, a time when Autumn and Summer sometimes join forces to tussle with Winter. Summer-like, extra warm days has everyone in shirtsleeves and then, within a day or two, a countryman finds comfort in a winter jacket and gloves whenever he is outside.

Windy, cold, raw days intersperse with foggy mornings and mellow afternoons. Sunny days with blue skies and a nip in the air are exchanged for dark, dismal, drizzly ones.

Hawks spiral overhead on thermal updrafts when the days are sunny and warm. This is the time of the year for them to migrate in huge numbers to the southland. Some few, like the red-tails and the accipiters and a few migrants from northern climes, spend the winter and haunt the fields and hedgerows and the roadsides.

One of the local young red-tails tries to out-maneuver an obnoxious crow that has nothing better to do than harass the young hawk. The hawk moves off to the south-east and the crow looks elsewhere for another diversion.

Two accipiter hawks flap their wings a few beats then sail a ways, and repeat this typical flight movement until they are soon out of sight.

Windy days are becoming more common as cold northern air clashes with the warm southern air. When the wind is out of the north it provides the migrating hawks with a tail wind that boosts their flight and increases their rate of travel.

Stiff breezes catch milkweed seeds away from their natal nests and fling than up forty feet or more where they are captured by the wind which then transports them out of sight. Winged seeds from a nearby tree flutter, spin and drift earthward as the wind moans through the bare limbs and branches of maples, walnuts and willows.

Chocolate brown cattails have gone to seed and their firm roundness has been transformed into over-sized cylinders of tan fluff.

Walking just inside a woods-edge, a countryman's thoughts are abruptly interrupted by a whir of wings. Thinking it to be a partridge, he is surprised and pleased to discover, upon a closer look, that it is a timber doodle, or woodcock. The chunky little brown bird in its camouflage suit had been probing in the damp, leaf-littered earth with its long, sensitive bill for earthworms. These cousins of the snipe have become scarce over much of its former range the past several years, so it was, indeed, a pleasant surprise to flush the little fellow.

Chickadees converse good-naturedly as the little acrobats cling up-side-down from twigs, inspecting them for anything edible. A countryman thinks that the chickadees, nuthatches and titmice are some of the most cheerful and friendliest folk encountered on an autumn day. They are pleasant company any time of the year for that matter.

Homeward bound, a countryman hears, in the distance, the familiar "kuk, kuk, kuk, kuk-kuk, kuk-kuk, kuk" call of the crow-sized pileated woodpecker. In his mind's eye he can see the large black and white bird with the flaming red crest, searching out destructive carpenter ants. Although it may excavate large oblong cavities in living trees, what might appear to be destructive may well prove the salvation of the tree. If the ants are discovered in time, the pileated may have just given that tree a new lease on life. It is, indeed, a true friend of the forest.

The many moods of November remind a countryman of the fickle ways of Mistress March. The winds blow wildly out of the west; sometimes they shriek down from the north. The fallen leaves swirl and eddy across the open spaces, piling up in protected corners, along old hedgerows, providing an insulating blanket and mulch for nearby plants and shrubs. Ultimately, they will provide nutrients to the soil, enriching it as they slowly decompose. Earthworms help hasten the decomposition process.

Misty mornings and afternoons of rain are not uncommon. Sometimes the rain steals across the land during the hours of darkness, like some stealthy feline, only to retreat to its hidden lair before dawn arrives. Then it is that a countryman awakes to a world filled with rain droplets hanging from every twig and stalk, each minute globule glistening in the morning sun, their iridescence creating a wonderland of liquid gems.

Wild geese hurry southward away from the cold blasts from the north country, eager to reach their warm wintering grounds. In March, they are eager to reach their nesting grounds even before the snow and ice has completely disappeared.

Also, like the vernal month, there may be calm, mild days when the occasional cricket will once more chirp and the sweet voice of a bluebird will be heard. Flickers call from the hedgerow and a jay scolds a gray squirrel because the squirrel happens to be in the area. The nutcracker pays it no heed and continues about its business.

The evening that follows remains relatively warm and a bat ventures forth into the gathering dusk one more time before retiring to a secure, draft-free abode to pass the coming cold months in the slumber of hibernation.

Following the warm, tranquil day may well be one that is once more fierce and blustering. Scattered snowflakes appear in the air and drift softly and silently earthward. The wind picks up, the flakes thicken, and soon the land is veiled with white.

The wind whistles and moans through the bare limbs and branches. It rattles the rust-colored oak leaves and drives the snow wildly before it.

The boughs of the spruces and firs reach out as though to catch the swiftly falling flakes. Crystal after crystal settle on the outstretched boughs until they take on a frosted appearance.

However, the storm doesn't last too long. It is but one of many short preludes to the ones that, all too soon, will come in earnest to cloak the earth with ermine.

Rays of the setting sun filter through the clouds. Several small wedges of ducks become silhouetted against the orange sky, their quacks conversational in tone.

A fork-horn buck stands outlined against the vineyard road. He remains motionless, a statuesque figure in the dusk, while several shots ring out in the distance. November is a month of change and even the deer senses that there will soon be changes for him also.

This is the time of the year when cold days and colder nights far out-number the ones that may be warm. Many mornings now find pools of water and puddles skimmed over with an icy sheath. There are afternoons when, on the north side of hills or ridges, a thin band of ice remains along the borders of streams. When the nights become cold enough to freeze the surface of the ground, these same sheltered spots will remain frozen while those exposed to the weak rays of the sun will thaw out during the day.

The first snows come in scattered flakes and brief squalls as the temperature fluctuates, but the time comes when warm, moisture-laden air collides with cold arctic air and the snow falls steadily. If the ground is frozen when this occurs there is a good chance that some of it at least will remain until the warmer days of spring arrive. On the other hand, the ground may not be frozen and the snow, once it starts to stick, acts as an insulating blanket. The warmth thus stored beneath the snow has a tendency to aid the sun in melting it at a later time. Eventually, the warmth of the earth is absorbed by the snow or else is later frozen, and the snow comes to stay.

Tracks in the snow can portray many events to one versed in the ways of the wild folk. Sparrows and finches hop and flutter about weed stalks as they feed, leaving not only footprints in the snow but prints of wing-tips as well. Mice leave tracks that appear like dainty stitches at the borders of the blanket of snow. Shrews tunnel beneath the snow, come up briefly to run along the surface, then dive below again in their constant search for food.

A red-bellied woodpecker gives its churring call as it clings to a grape vine that has gone wild, it spends several minutes feasting on the frozen fruit.

Cedar waxwings fly about in small flocks in search of berries and other small fruit. Waxwings are gregarious and also mate for life, each one very attentive to its mate. A countryman watches two as they pass a berry back and forth for several minutes, each politely refusing to eat the proffered morsel, until it is finally accepted by one of the birds. Each time he witnesses an event like this he contemplates their apparent tenderness towards one another. We of a higher order of creation could learn many valuable lessons from some of our wild neighbors.

During one of the snow squalls, a pileated woodpecker slowly makes its way from one woodlot to another. The fragmented and cut over woods do not provide the environment needed by these crow-sized birds, and a countryman is always glad when he is fortunate to hear or see one.

As the flakes continue to profusely fall, a red-tailed hawk soars overhead looking for a hapless squirrel or unfortunate mouse. A small cottontail would do just fine on this stormy day.

Cold and stormy weather often bring visitors from the north country in the form of evening grosbeaks. These chunky birds are slightly smaller than robins and wear suits of soft browns and tans with touches of yellow here and there. They are noisy and fly about in flocks until they find a bird feeder, whereupon they descend upon it and in no time they have shelled any sunflower seeds with their massive beaks. Those beaks were designed to separate seeds from the cones of the conifers in the northland. When food there becomes scarce they drift southward until food is found. When they appear at the feeders, one knows that the blasts of winter are not far behind.

When the snow covers the land it creates, as it were, a fresh, clean slate upon which can be written the comings and goings and all the other activities of the wild creatures. A countryman will learn many things, such as who is about, what they did or what may have befallen than. Many such activities occur under cover of darkness while others take place during the hours of daylight.

When the wind sweeps across the land it often wipes the slate clean as also does a light, new fall of snow.

Sometimes the snow will lie deep upon the land, so much so that one cannot easily move about. The larger wild folk also have a more difficult time while the smaller ones can move effortlessly across the smooth, soft surface.

At times like these, a countryman listens even more intently so as to hear what he often cannot see.

A red-bellied and a pileated woodpecker each call as they forage in different portions of the woods. White-breasted nuthatches call in a nasal tone to one another as they clamber over and around branches and up and down tree trunks. Titmice call back and forth to one another as they go about their business. House finches and tree sparrows chirp and twitter companionably from the thickets and weed patches. A crow and a blue jay join forces to call accusingly after orange-clad hunters slowly and stealthily walking along the edge of a field.

The sun tries valiantly to warm the late autumn morning and does succeed in inducing several bluebirds and a robin to softly warble and chirrup as they relish the subtle warmth. Perched quietly on a power line, a mourning dove sadly surveys a cornfield where it has been finding sustenance. Now its food supply is covered by the heavy, deep snow.

A sharp call alerts a countryman to yet another bird. A flashing of white with gray betrays the flight of a mocking-bird as it flies from bush to bush, pausing at last to feed on frost-softened berries.

Although usually a vociferous summer resident, the mockers were conspicuous by their absence during the past summer, and a countryman is glad to hear and see one of them again in the area.

This is the season of thanksgiving, a time to reflect on the harvest of many varied crops, of being able to store away the necessities for the long, cold months ahead. This is an especially auspicious time to give thanks to the Creator of all things for His care and for all the wonderful blessings He has bestowed during the year.

A countryman is thankful for all this and also for eyes and ears to see and hear all the wonderful world about him. He is thankful for times of quiet, yes even silence when there is nothing to disturb one's innermost thoughts. He is grateful that he no longer lives amongst the throb and throng of the city.

This is, indeed, a time for heartfelt gratitude for friends for family and for all the many good things we enjoy in this great land of ours. Let us always remember to be thankful for the little things of life that we so often take for granted.

The countryside is relatively quiet these days. For the most part, migrations are over and only the winter residents remain. Harvesting is finished, chipmunks, woodchucks and bats have settled in for their winter sleep. Others that may nap later on are still up and about.

A crow caws hoarsely as it pesters a mature red-tailed hawk. The hawk soars a little higher trying to ignore the black pest as it scans the ground below for a meal. The crow is persistent but the hawk makes no protest, continuing to soar in ever-widening circles.

Suddenly there is a protesting, shrill whistle that sounds like the hawk but comes from a clump of sumacs. A countryman turns towards the sound and is not surprised to see a blue jay calmly eating the burgundy fruit.

Blue jays are noted for their accurate mimicry as well as creating disturbances. They will call like a predator then watch the little folk either freeze, motionless, where they are or scamper to a secure hiding place. One can almost see the jay smile at this; sometimes there is an audible sound that appears suspiciously like a snicker.

The nesting season is long over and there is no need to squabble over territories, so the cardinals gather in flocks of several pairs to rove about the countryside. They seem to enjoy each other's company and will continue to until late winter when lengthening days tell them that it is time to think about courtship and territories once more.

Fox tracks meander across a field and along a hedgerow until they intermingle with those of several cottontails. Further on, several tufts of soft, brown fur bear witness to an encounter between the two species. The fox may have gotten a meal or perhaps the rabbit escaped with a couple of tears in its coat.

Flying low overhead, a ring-billed gull sets its wings into full soar and casually scratches the side of its head with a foot, then continues on its way. One does not think of birds doing something like that while in flight, but they are at home in the air and may perform many acts of necessity quite spontaneously and with little or no effort.

The big pileated woodpecker, in the old ash tree with the dead top, was looking for food. So was the Cooper's hawk that lit on a limb nearby. Both birds were nearly of a size and the hawk did not seem to make an impression on the woodpecker as it hitched up a large limb, chipping off pieces of bark as it went.

The hawk flew out a few feet then soared swiftly in at the woodpecker. The scarlet-crested bird continued to forage, dodging around the limb it was on to avoid the onslaught of the hawk. The hawk circled and dove several more times. Frustrated, the pileated flew several yards out from the tree and returned to continue its search for food. The hawk flew about, perched beside the woodpecker, flew out again and made several more dives before it gave up and headed out across the vineyard in search of more suitable and smaller prey. A few moments later, the big woodpecker left the ash tree and flew slowly and steadily towards the nearby woods.

Later, a countryman heard what sounded suspiciously like laughter coming from that same woodlot.

The tattered, dry leaves of the cat-tails rustle and whirr in the sharp, afternoon wind. Clouds scuttle across the sky as if to escape to a warmer clime.

Tree sparrows with their rusty caps and dark "stick-pin" in the center of their chests have arrived for the winter and are busily seeking seeds from the weed stalks. Some search the ground for seeds already fallen.

A countryman enjoys listening to the cheerful chatter of the tree sparrows, and likes to think of them as miniature chickens conversing amongst themselves as they pick and peck on and below the weed stems.

Listening and watching, a man detects a darker bird feeding with them on the ground. Looking more closely, he can see that it is a song sparrow that has decided to spend the winter, as some of them do. When the weather becomes colder and more severe, they will spend most of their time in heavily tangled ravines or thick, overgrown hedgerows. Southern exposures are favored spots for them as well as other species. The weak, winter sun can more readily warm the south slopes and they are protected from the icy northern blasts.

Moving slowly away, a man hears a fragment of melody from the little striped songster and he silently wishes him health in the days ahead.

While the year draws steadily to a close, a countryman is quite apt to review the events that have occurred during the past months, the subtle changes from bud to fruition and, finally, the harvest and the frost.

Where in spring the strawberries blossomed and, later, the raspberries and blackberries were succulent and sweet, he now finds only traces of their existence. Frost and wind, cold rain and driving snow have already taken their toll. The canes stand exposed and naked, stripped of their leaves, but they are well able to survive the cold.

The bluish-white canes already bow towards the earth and winter snows will assist them in reaching their goal and by doing so they will take root to form new plants, thus propagating the species.

Blackberry canes stand sturdy and regal; in many places they form thorny, impenetrable thickets that Peter Rabbit fondly referred to as the "Dear Old Briar Patch", indeed, they do provide safe, secure shelter for many of the wild folk in times of danger. Hawks, owls and other large predators cannot penetrate their tangled, thorny depths.

Along a road-edge, a deer lies quietly in death, its life-purpose ended. Yet, even in death, it continues to serve the needs of others. Passing by in his vehicle, a man notices the hunched-over form of a red-tailed hawk gaining strength and sustaining its life-forces from the death of the deer.

Others, like foxes, coyotes, bobcats and smaller creatures, will benefit also from its death. Life and death, so closely intertwined.

A countryman thinks that this should be true in our lives as well. If our lives are full of loving, caring and sharing we can touch the lives of others in a meaningful way during our brief sojourn here. Conversely, we will continue to exert a positive influence on those we have left behind, in their memories of us, long after our demise.

May we so order our lives that when our days draw to a close we will leave behind a beautiful sunset.

# WINTER

The great orbs of the astral clock are moving steadily towards the shortest day of the year. The sun, like a great glowing pointer on the clock's celestial face, is setting nearly as far southwest as it will. Soon it will begin its northward journey, inexorably moving a little more each day, and gaining momentum as the weeks pass.

There are years when winter cannot seem to wait until the official time of its arrival. Sometimes a frigid air mass moves down out of the arctic and temperatures tumble quickly. At times like these, a countryman will wonder if it will be a long, hard winter or, since it is getting an early start, perhaps there might be an early spring. Since he is not one to prognosticate the weather, he must be content to wait and see what unfolds in the months ahead.

When there is only a small amount of snow on the ground and the weather warms a bit, the wild folk take advantage of this and move about more actively.

Such a break in the weather caused a skunk to stretch, perhaps yawn sleepily, before leaving the old woodchuck burrow for a short excursion, leaving dirty tracks as it wandered about. The snow soon cleaned off his feet, no doubt leaving them their normal rosy color, further enhanced by the cold snow.

Taking long strides, a wild turkey leaves large, chicken-like tracks behind as it moves purposefully through a vineyard, across a roadway and into the woods.

Cottontails, squirrels and deer also have been about. The wide-splayed hoof prints of a large buck gives assurance that he survived the hunting season, but he has yet the winter to endure. Winter weeds out the old, the weak, the sick or injured, and also many of the young from among all the wild ones. Predators do their part in alleviating suffering as well as assuring that most that survive the rigors of winter are those most fit to perpetuate the species.

A hunting fox leaves its almost in-line trail in the scant snow cover as it searches out meadow mice or some other small creatures for a meal. In years past, a countryman has followed a fox for a mile or more, and was amazed at all the things the fox did, the animals or birds it had encountered along the way. One learns much about the outdoor world and those who inhabit it by following trails or just by standing or sitting quietly for some time.

Christmas arrives shortly after the Winter Solstice and what a joyous time it is, with family gatherings and their attendant excitement and good cheer of the holidays.

It is the season when wonderful odors come from kitchens as various kinds of cookies, pies and candies are made.

Then there is the tree, standing in its place, all shining and shimmering with presents tucked away underneath.

This is also a time to reflect on the past. A countryman likes to call to mind that night long ago when the Christ-child was born and lay in a manger in an animal shed. Although there was no room for Him at the inn, the animals were willing to share their humble shelter with their Creator.

The shepherds came, and the wise men bearing their gifts, came to worship Him whose Star was seen in the East. The greatest of all gifts lay in the manger, but few recognized its worth.

Our best gifts are what we do for and how we consider those less fortunate than we are, no matter what our station in life may be.

The lightly crusted snow crunches underfoot as a countryman walks along the edge of the old field. Others have been here before him as he notices the tracks of deer and also those of a dog. He is pleased to note that this time the two sets of tracks are going in opposite directions.

A downy woodpecker whacks noisily on a sumac branch while several house finches perch in adjacent branches. The males wear suits of brown with rose-colored vests and touches of rose about the head while the females are dressed in grayish-brown with darker streakings on the chest and along the sides.

Tree sparrows hop about the crusted snow among the dried remains of asters, goldenrods, other weeds and various grasses that reach above the snow on a now-dormant field. Occasionally, one will jump up to reach a likely-looking seed head, sometimes perching on a stalk while it feeds.

A group of white oak trees seem to huddle together as they stand watch at the edge of the woods. Their leaves, now dried and brown, cling to the branches, refusing to be released. They rattle and rustle with each passing breeze, seeming to carry on a conversation all their own.

Winter has officially arrived and the land waits, almost seeming to hold its breath, for the cold and snow to finally take it firmly within its grasp.

The temperature hovers just below freezing and an occasional flake of snow spirals slowly earthward.

A small band of chickadees work their way through the woods and a pair of titmice follow in their wake, their nasal "dee-dee-dees" contrasting with the clear, sprightly calls of their chickadee cousins. A nuthatch adds its "yank-yank-

yank" call and for a few minutes the woodland seems replete with life. Then just as suddenly, it is quiet once more.

Several bluebirds, previously unnoticed, take wing from low bushes in the field. Soft, sweet but somewhat sad notes are uttered as they depart. A countryman ponders the fate of these gentle creatures that seem determined to share the winter with him. He trusts that they will survive to announce the end of winter just as they now have announced its arrival.

Upon entering the woods, a man follows a deer trail for a short distance until it goes in a direction he chooses not to follow.

Soon he is surrounded by wild turkey tracks that meander here and there, under low-hanging branches, up over fallen limbs, sometimes separating for a ways, then rejoining as the flock forages its way through the woods. There are only a very few scratches in the snow, indicating that the flock is searching in vain for food that never developed on the overhanging branches in the first place.

Back-tracking the flock, a man comes to a group of tall pines deeper in the woodlot. They have provided a sheltered roost the previous night, made evident by the droppings scattered over the snow beneath.

Homeward bound, a countryman feels empathy for the wild ones who must endure the harsh months ahead. He knows that just as surely as the old year is ebbing away that when Spring finally arrives once more, many of the wild creatures will still be here to welcome the warming days.

December, like the last leaf on a tree, relaxes its grip on the old year and takes its place with the other months preceding it.

The last month has seen the final holiday of the year come and go. The shortest day quietly, yet surely, arrives bringing with it short days and long, winter evenings.

A countryman well remembers his winters in Maine when, for several weeks, he would leave home for work in the pre-dawn darkness and before he would arrive back home, darkness once more had settled over the land.

Many the early evening he had shouldered his double-bitted axe and ventured forth into the woods under a brilliant full moon. The light, reflected from the snow, made wood cutting easy and pleasant and extended the short day. Walking back on the narrow path with a birch log upon his shoulder, a man could relish the crisp, cold air and the mellow moonlight.

When January slips around a drifted shoulder of the land, there is a sense of newness in the air. A new season already has begun and the new year holds a promise of new beginnings and a fresh start in life.

As a man looks back over the year that has just drawn to a close, he contemplates what this fresh new year will reveal. He wonders what discoveries he may make as he turns each leaf of Nature's book in the weeks and months that lie ahead. He likes to think of each new year as another year of school and the natural world as the Creator's classroom.

At the beginning of each year, a countryman will start a list of birds that he sees.

Winter is an excellent time for anyone to start learning about the feathered folk if they so desire. Only the more hardy creatures are out and around and there are relatively few winter residents and the few migrants that casually stray south add excitement as one tries to discover who they might be.

A typical list will usually start with crow, blue jay, chickadee, and nuthatch. Some years it might start with a woodpecker or house sparrow. A wintering robin or bluebird sometimes can be found. Soon one has a nice list started and before the spring migrants start arriving one has become more confident in his or her observations.

An inexpensive field guide with good illustrations of the more common birds can be a great help. So can the local library. Later on, if one really becomes "hooked" on birds or the outdoors in general, quality field guides on many nature subjects are available. One no doubt will want a pair of binoculars to aid in identifying these creatures that do not like to have one come too close to them, or persist in staying in the tops of the trees.

Seeing is an art to be learned. Many people look at the world around them and fail to see what they really look at. A countryman thinks this is partly due to the many unnatural things that fill our world today. We are used to the artificial, the man-made, and the mechanical. We tend to be myopic when it comes to the natural world. But this can change, with effort and practice.

Just start with your dooryard, your street or perhaps the vacant lot if you live in town or an urban environment. There is more going on than one would believe possible. When you have learned to really "see," then the natural world will reveal more and more to you and you will find a sense of calmness and peace that you may have never known before.

There is no telling where your rambles will take you this new year.

January days are usually cold and windy with snow squalls driving down out of the north. Winter has settled itself comfortably over the land, assuring the world that it will be around for a goodly number of weeks to come.

Sometime during the month there occurs what is generally know as the January Thaw. A countryman looks forward to that time when the snow softens in the sun's warming rays. Icicles form as snow becomes liquid and refreezes in the cooler air. The water flows over the surface of the icicle, freezing in thin layers as it flows downward to drip off the ever-elongating end. This flowing and freezing action causes the icicle to continually thicken at its base while ever growing longer as small amounts of water continually freeze at the tip before dropping to the ground.

Hanging in rows from the eaves, the icicles glisten like so many silver daggers in the sunlight. As the sun disappears at the end of the day, the snow ceases to melt, and finally the last little drop hangs and quivers ever so slightly at the tip of the icicle. If the air has cooled quickly enough, one can watch the droplet as it solidifies and becomes a part of the whole.

The thaw lasts several days or several hours, or may not arrive at all, waiting until February to put in its appearance. Whenever it does arrive, it is welcomed by both man and beast.

Snowstorms pass over the land more frequently now, adding layer upon layer to the ermine blanket that now insulates the earth from the harsh winds that moan through the tree-tops, whistle around the eaves and pile fragile snowflakes into hard and beautifully sculpted snowdrifts.

Before the snow becomes too deep, a man likes to visit the little ravine, the brush lot, the hedgerow and several other places he knows are favored by the wild folk.

The snort of a deer comes from the brush lot. The doe stands motionless; her ears swiveling to catch each sound and evaluate it for any possibility of danger. She senses the presence of potential danger. Like a warning flag, her tail goes up, flashing white even as her body gathers momentum as she bounds away in great graceful leaps and soon disappears from sight.

A man continues until he comes to a shallow drainage ditch. Here, several deer have broken the ice so as to reach the clear, cold water beneath. The lightly congealed water splashed over the snow reveals to a countryman that it must have been only a few hours ago that the deer had broken the ice for a refreshing drink.

Farther up the ditch is a small pool where the current keeps it from freezing. Both sides are imprinted with numerous bird tracks as a flock of crows recently have gathered to slake their thirst.

The hedgerow forms a protected lane of travel for squirrels and cotton-tails. In more overgrown portions, it provides sheltered resting and feeding places for the small mammals and various birds.

A cardinal moves quietly through this protective cover, pausing here and there to sample frozen berries. They seem to be to his liking as he feeds for several minutes before moving on.

Trudging homeward through the snow, a countryman hears the familiar voices of a rollicking group of chickadees, a hairy woodpecker and the "clee-up" call of an over-wintering flicker. Familiar with the many and varied sounds of nature, he does not need to see the owners of these voices to know who is out and about. It is enough just to hear them and he is content to enjoy their many tonal qualities.

The snow sifts steadily downward adding layer after layer to the already thick white blanket covering the bosom of the land.

A gray squirrel sits in the sturdy crotch of a limb on a maple tree, gnawing open a black walnut. His teeth rasp against the hard shell of the nut, the sound carrying some distance in the still, crisp air. Already his gray coat is covered with snow, the guard hairs holding the flakes away from the warmth of his under-fur thus keeping him dry and warm. Flakes pile up on his head between his ears while his bushy tail, curled over his back, catches the snow and warms his backsides.

In a tree beside a roadway sit two red-tailed hawks just a few yards apart. Their feathers are fluffed out providing the warmth of a down-filled jacket while the short, soft feathers on their legs and thighs give added protection from the cold as they patiently wait and watch for prey.

Winter hawks must be hardy as well as adaptable to weather and hunting conditions. Most of the small mammals now travel and forage beneath the snow and out of sight. Cottontails tend to stay close to thick cover of bushes or bramble patches. Occasionally a careless or unwary squirrel will provide a meal.

The boughs of the pines and hemlocks and the spruce and fir plantations bend beneath their burden of snow. Their needles catch and hold the snow and beneath their drooping branches, the ground remains bare and relatively dry for some distance away from the trunks of the trees.

Many years ago, a countryman discovered protection from the elements afforded by the conifers and he also learned that many of the wild folk utilize the shelter also, not only in winter but as protection from wind and rain in the warmer months.

Grouse and turkey roost among the branches during the night, selecting a spot on a branch close to the trunk of the tree. Owls may spend the daylight hours in the same trees. Droppings and pellets of regurgitated fur and bones will indicate the species which roost in the branches above.

If a landowner is fortunate enough to have a grove of conifers in his woodlot, he many expect to hear the tiny tin horn call of redbreasted nuthatches or the chirring and ratchet-like vocalizations of feisty, rambunctious red squirrels.

Early one morning, a countryman flushed several mourning doves from a large spruce at the edge of the village where they had spent the previous night. The thick, overlapping boughs provide protection from predators and wind and cold.

The same morning a mockingbird was seen in a bush, its feathers were fluffed out to provide warmth and it was sitting on its feet, keeping them warm as it surveyed the surrounding area. Several minutes passed, and then it was joined by another, no doubt its mate.

Mockingbirds, red-bellied woodpeckers, even titmice and cardinals were not always residents this far north except perhaps near the warmer Atlantic coast. Over the years, the climate has changed, more bird feeders have been put up and these species have expanded their range. There also have been changes in habitat over the years.

As January reaches its midway point, Old Man Winter settles in for the season. The wild folk who do not spend the cold months sleeping or at least napping, have adapted their lives to the weather and shortage of food.

Deer and grouse now are eating buds, the deer browsing the tender tips of branches as well as the buds. As the snow deepens, the deer will tend to congregate in "yards," where there is sufficient browse and where their hooves have compacted the snow into trails and they are able to move about easily and conserve as much energy as possible.

The winter days are becoming noticeably longer at sunset time when the orange orb delays its descent by at least a minute each day as it commences to swing northward once again.

The celestial clock also has started to lengthen the day at its beginning as well, the sun now rising a minute earlier each morning.

Although the winter days may often be gray and gloomy, yet the hours of daylight lengthen more and more each day. The great cosmic clock ticks steadily, inexorably away as seasons progress with their resulting changes of climate and weather patterns.

Dawn breaks over the eastern ridges on a crisp winter morn. As the rim of the sun slowly slips above the horizon, shafts of light stream upward and the morning sky is suffused with shades of orange interspersed with various shades of yellow.

A countryman watches, immersing himself in the atmospheric glory of another new day. Each new day is different; sometimes there are only subtle variances, but even these are easily detected by one who is in tune with the rhythms of Nature.

A light snow has fallen overnight and now each fresh, new crystal responds in sparkling splendor to the bright rays of the rising sun.

Sometime, between snowfall and day-dawning, some small animal has made myriad tunnels just beneath the surface of the snow. The tiny, snow-arched tunnels create complex designs over the surface of a small field.

Upon closer examination, a countryman determines that all of this energetic work is the work of a shrew, a small gray, mouse-like creature.

The shrew is no mouse, however, but will dine on mouse or vole when the beetles, other insects and earthworms upon which it normally preys are unavailable. This animal, no doubt, was in search of voles.

A shrew's system is such that it almost continually must hunt for food to replace all the energy that it is expending. To watch one of these diminutive creatures, one will wonder when they find time to sleep or perform other functions of life. This is another of Creation's wonders.

By now the sun is well above the horizon and its warmth can be felt even though its rays yet slant in at a shallow angle. Overhead, a half-moon glows faintly against a cerulean sky.

Crows caw in the distance, tree sparrows and house finches twitter and chirp from the nearby thickets, and another day has commenced.

During the course of the winter, weather fronts merge, override or collide. Warm air from the South pushes in and rain develops. The temperature rises rapidly and the dry, fluffy snow becomes filled with moisture and water starts flowing from the snow pack.

Under the combined onslaught of warm air and rain, the snow rapidly shrinks until the white blanket disappears from the land except for those places where it drifted and compacted.

Then a cold front arrives, the warmth retreats, and the temperature plunges into the single numbers once more. What had been liquid now changes its form in several dramatic ways. Standing water becomes ice, icicles freeze to a standstill and water vapor crystallizes.

A crust forms over the now hardened snow and the rain already has changed back to snow which forms a thin coating over the landscape.

Brooklets and streams had opened up during the thaw. Thin sheets of ice reach out from the snow banks on either side of the flowing water until the turbulence near the center keeps it from freezing. Beautiful and intricate patterns develop as the surface of the water crystallizes.

Delicate and fragile crystals form on twigs and weed stalks overhanging the flowing water. Twigs that are occasionally splashed by water become encased in shining silvery sheaths of ice. Small, bell-like ice formations cling to the tips of branches that extend to the water's surface, growing larger as the hours pass and thin layers of water congeal on their surface in the frigid air.

Since most of the snow has gone, a countryman once more can explore outlying areas. The dusting of new snow on the remains of the old provides a fresh new page on which the wild folk can trace their identities and activities.

Across one snow patch, a meadow mouse or vole left dainty footprints as it headed towards a weed patch. Each tiny toe and foot-pad was clearly imprinted as it scurried along its way to concealment and safety.

The prints of a crow also clearly were displayed as it walked about and probed the hardened turf with its stout beak for something edible.

Nearing an open woodlot, a man sees the footprints of a pheasant as it has come out into the fallow field. Its tracks are chicken-like, not as large as those of a turkey, yet larger than those of a grouse. Other subtle differences are apparent to his eyes that distinguish between the other two fowl. Such is the art of tracking and seeing.

Entering the woodlot, a countryman notices the tracks of a large deer and discovers the oval bed in the leaves which it has left. The dusting of snow surrounding it shows that the deer had bedded down while it was snowing and had not left until the snow had ceased.

Movement near the top of some sumacs attracted the attention of a man, almost home. Focusing his binoculars on the movements, he discovered three pairs of bluebirds hungrily devouring the red fruitlets. It never ceases to amaze a man how these birds survive the winter. Many do not. Only the ones that can find sufficient food and warm shelter will survive.

Movements in a nearby tree in the hedgerow disclose two diminutive birds, smaller than chickadees. Again, the glasses reveal that they are golden-crowned kinglets, insect eating birds that are hardy little mites. Their plumage is a subdued mixture of grays and whites. Their crowns are striped with yellow and that of the male has an orange stripe in the center of his gold crown. They are, indeed, small kings of the hedgerows and woods. They prefer the thick cover of conifers, but hunger drives them wherever food may be found. A man's day is now complete.

Appreciably, the days lengthen as the first month slips away, giving place to the month of sweethearts and Valentines. Each day now, the day star rises earlier each dawn and sinks below the western horizon a little later each evening.

Arctic air flows southward, down from the realm of the great white bear, locking the land in its frigid embrace. The old adage still holds true as it has for many a year, that as the days begin to lengthen the cold begins to strengthen.

To the Native Americans, the full moon of February was aptly known as the Hunger Moon. Winter now reaches into its icy depths to cast its most chilling treasures across the landscape. The cold penetrates many feet into the ground in some places. This is the hardest time of the year for the wild ones. It is indeed the moon of hunger, which to some becomes the time of starvation. Only the fittest survive.

Those creatures that hibernate or spend the cold months in some form of dormancy are not adversely affected.

The woodchuck, or groundhog, proverbial prognosticator of the return of Spring, often as not sleeps right through his official duties at the first of the month, preferring to remain curled up in his snug nest in his burrow until a warm spell arrives later in the month. By that time, other urges take precedence over prognostications. His biological clock awakens him to bestir himself and seek a mate for the coming season.

The woodchuck may visit several neighboring burrows, leaving his muddy footprints in the snow as he travels abroad. For now, though, he will slumber on, oblivious to the extremes of temperature and the knife-like winds that howl a few feet above him.

The bat, in its hollow tree cavity or in a cave, sleeps peacefully the wild winter weather away. Its body temperature has lowered and its pulse down to a minimum while its breathing has diminished in proportion.

Meanwhile, a female hairy woodpecker is successful in dislodging hibernating insects or perhaps some in the larval stage from beneath the bark of a decaying branch. She stops her activity and freezes motionless as a sharp-shinned hawk flies through the woods, dodging in and out between the trees. As the hawk disappears, she continues to forage.

The light snow cover reveals tracks of a red squirrel that has been running about in search of hidden caches of food. A countryman is pleased to note that there is one of the little red rascals in the area. This is the first evidence he has had of one in several years.

Rambling on to an adjacent woodlot, a man spies something reddish-brown on a tree branch. At first glance, one would think perhaps it was a cluster of leaves attached to the limb.

To a practiced eye, something is not quite right for leaves and the binoculars once again are put to good use.

Ah—it is the little red rogue himself, or herself, as the case may be, sitting there with a small walnut watching me as much as I am watching it. A few moments pass. Down the tree trunk he flows and across the snow, his tail held at a jaunty angle as he bounces along toward the security of a large white pine. It is good, a man thinks, to be able to see one of these feisty little fellows once again.

A few more weeks and Mistress Spring will be starting North once more and a countryman indulges himself in these pleasant thoughts as he makes his way homeward, his boots crunching in the snow.

When a severe cold wave sweeps over the countryside, it will drive even the hardiest of birds to the shelter of deep woods. Lacking this type of retreat, they will seek refuge in the recesses of deep, wooded ravines and glens. Dense thickets with evergreens are also sought out.

Many animals seek the same types of cover in order to escape the harsh, howling winds and the biting cold. Conifer thickets are sought by deer as relief from the elements. Turkeys, grouse and other species seek security and protection beneath the overhanging branches of large hemlocks and pines. Predators and prey alike suffer during the extreme cold.

Most of the common animals that are out and about during the winter months are active at night for the most part. Only the squirrels are to be commonly seen during the daylight hours. On occasion, a rabbit or a deer will be flushed from their place of concealment. The fields and open woodlots are almost barren of wildlife. Even the usual woodpeckers and nuthatches are conspicuous by their absence. Only the sound of the the wind is heard as it moans through the tops of the maples, oaks and ashes. The wind sighs soulfully as it bends the branches of the hemlocks and twists the limbs of the pine. A crow sits huddled over its feet, its feathers fluffed against the cold.

A skein of ducks is silhouetted against the leaden sky as they move from one lake to another. A sharp-shinned hawk alternately flaps and sails at tree-top level, then ducks into the woods and dodges among the tree trunks in an attempt to flush out a meal.

Walking about the seemingly lifeless countryside, a man is bound to notice evidences of continuing life and even the hint of new life.

Feeding under the cover of darkness, cottontails have gnawed the bark from last year's new shoots of sumac. Standing on their hind legs, they nearly stripped the bark from the shoots to a height of two feet or more. Of course, some of this was done when there was still more snow on the ground which aided them to reach higher. The tender inner bark of blackberry canes is also a favored food as is that of small trees and certain small shrubs.

The soft, fuzzy, gray-green leaves of the great mullein hug the ground, waiting patiently for the warmth of spring to stimulate growth once more.

Small willows grow along a damp ditch and along their outer branches bits of white show in evenly spaced rows. A countryman knows what they are, but cannot believe his eyes. Upon closer examination, he sees tiny, soft furry forms have started to emerge from their protective sheaths. Diminutive they may be, but pussy willows they are, bravely peeping out at the hostile environment.

Here is life, in all of its newness, patiently waiting as is the mullein, the primrose, the phlox and all the other perennials, for the sun and warm southern breezes to warm the air and the ground once more so that they can burst forth into renewed growth and bloom.

For now, all of nature rests and waits for the great wheel of the seasons to turn in its inexorable course, bringing with it warmth and strength and nurture.

November and February have a lot in common. The capricious month of November keeps the front door open to Autumn while Winter repeatedly knocks on the back door and often is allowed in for a brief visit. So it is with February. Although Winter appears to hold eminent domain over the countryside, yet fickle February will open the front door to see if Spring would like to enter. On occasion, a mild southern breeze may take advantage of that open door and venture inside for a transitory visit and a day or so of mild weather will be relished by all concerned.

Sometimes the warmth and cold will clash in such a way that the land becomes enveloped with a thin veil of fog and the whole landscape will take on an ethereal quality.

Perspectives change; the whole atmosphere is subdued and all harsh outlines are softened. Wisps of fog drift among the evergreens, its tendrils curling around the branches while in the woodlot it swirls around the trunks of maples, oaks, hickories and walnuts.

Silently, a grey squirrel glides over the branches of a maple, appearing more of an apparition than a reality, its form blending and flowing with the fog.

A countryman enjoys a walk on a foggy afternoon, out through the vineyard, past the spruce grove, past the woodlot, through the hedgerow and along the edge of the large open field.

The grove of spruces always entices a man to stop and look and listen intently for a minute or so. This day is no exception, and as he lets his gaze wander about, he notices a blot of blue amongst the green boughs of one of the spruces. A blue jay, feathers fluffed against the dampness, sits quietly on a branch. Bird and man observe one another for a few moments, and then the man slips silently away. The jay remains unperturbed and motionless on its perch in the spruce.

Two crows converse in the distance. Then another voice is heard, a hoarse, croaking voice, now coming from overhead.

The voice is recognized as that of a raven, and looking up, a man sees the large black bird with its wedge-shaped tail.

Alternately flapping and gliding on flattened wings, it soon disappears from sight. One last hoarse call floats back to a man's ears, reminding him of the many ravens he heard and saw during the years he spent in the north country of New England.

Seeing or hearing the unexpected is one of the many rewards of being out-of-doors.

The sun slides steadily westward towards the distant ridges between the lakes. A cold, raw wind starts up from the northwest and the fog dissipates. A man turns homeward in a contemplative frame of mind, feeling enriched by the sights and sounds of the squirrel, the jay, the crows and the raven. And surrounding it all, the nebulosity of the fog.

February, the month of hunger and fickleness, is also the month when the fire burns its lowest on the hearth of the land.

An inch or so of new snow overnight creates a new page on Nature's sketchpad. Crows have created cross-stitch patterns in several places as they have wandered about looking for voles or some dormant beetle hidden away in a grass tussock.

Gray squirrels were up and about as the first light broke out across the eastern ridges, long before the sun showed its shining face. Earlier yet, a fox had moved purposefully across the roadway and into the nearby vineyard and back again and into the adjacent woodlot. At first glance one would assume that one of the resident woodchucks had awakened and gone about for a late winter stroll. After all, it is that time of year. However, upon closer examination of the tracks, they prove to belong to someone wearing a black and white coat. A skunk, or maybe two, has been exploring and foraging for food, taking advantage of the warm spell. On the other hand, February is the courtship month for the skunk tribe. This may, in fact, be an even better explanation for all the comings and goings about this particular burrow.

Later on, a countryman came across more skunk tracks and at one point two animals were apparently traveling together. After all, this is the month of valentines and love.

The strong winds and the heavy wet snows that came earlier in the season damaged many trees, large and small. Branches and large limbs were broken off and some trees have twisted and broken trunks.

An old apple tree, rotted and partially hollow in the center, was broken off several feet above the ground, its top branches on one side resting on the ground. No longer will it bear apples for the wild folk to feast on, yet in its death its branches with their tender bark has provided nourishment for cottontails during the winter. Nothing is wasted in nature; everything is utilized, everything has its purpose in the scheme of things.

Winter winds also destroyed a freshly made woodpecker home hollowed out of a dead willow branch. The carefully constructed apartment was broken off several inches below the entrance hole, and lay desolate on the ground beneath the tree. One would hope that the unfortunate bird was not inside when disaster struck, or that it was not during the night.

The sun broke out through the cumulus clouds, warming the earth. At least eight bluebirds perched here and there on sturdy weed stalks and briar canes. Several were seen characteristically fluttering their wings while several others flew to the ground, apparently seeing something edible. That particular area of fallow field seemed alive with bluebirds for many minutes until they moved on to another locality.

If they can hold out for a few more weeks, warmer days and nights will make their little lives much easier.

For now, they must take each day as it comes and make the best of it. Soon the cold will be gone and their sweet warbles will once more be heard across the warming land.

During the final days of February, the March Lion put in many long hours of practice, roaring about the countryside and creating mischief in many places. Then for a day or two he retired to some secluded place for a nap as felines are wont to do after exercising strenuously.

A countryman decided to ramble about a bit on one of those windy days, not expecting to see many of the wild folk about, but knowing that other interesting things might be seen. Then again, perhaps not.

Cumulus clouds raced across the sky like clipper ships before a gale, while their shadows chased one another swiftly over the meadows and fields in a way resembling children playing tag.

Two red-tailed hawks sailed confidently overhead, displaying their skills in utilizing the various wind currents.

Several crows were endeavoring to do the same but with much less success. At times they struggled to reach their destination and had some difficulty achieving a landing, while others, going with the wind, often found themselves overshooting their perch.

The two species readily showed their aptitude, or lack of it, in negotiating the sometimes violent air currents. Perhaps that is why the raptors have more respect than crows.

A small flock of house finches arose from their tree-top perch and allowed the wind to sweep them away to their destination, while a cardinal, desiring to go in a direction against the wind, flew low and just above the matted weeds where the air was less agitated until she reached the shelter of thick bushes.

All that night the wind blew until nearly dawn when it subsided to a light breeze. It had scoured the sky and swept it clean of every trace of clouds. The morning sun shone brightly against the azure sky and it slowly warmed the air.

A nuthatch laughs "ha-ha-ha-ha" as it walks headfirst down a hickory limb in search of breakfast while back in the woods a titmouse is calling earnestly for Peter. Hesitantly, a male cardinal calls "purty, purty, purty," perhaps realizing that it isn't quite time to start staking out his territory. But a fellow does need to practice a bit.

Turning homeward, a man hears a familiar voice away in the distance. He stops and listens intently for a few moments. He scarcely can believe his ears as a robin is cheerfully caroling. It is one thing to see a robin or even several robins, but to hear one caroling in February is remarkable.

A bit later one flies overhead and is soon followed by yet another. No doubt that one of them was the caroler as they were flying from the proper direction and that voice of springtime was silent.

In the distance the sweet warble of a bluebird is heard. Then the voices of red-winged blackbirds are heard; some call "chek, chek, chek," while others gurgle "o-ka-lee" as they move restlessly about the branches in the top of a distant tree.

A short time later the blackbirds have moved on, perhaps further north as they usually do when the first ones arrive.

A man ponders what he has heard and seen. Could Spring be coming early this year? Time will tell. Meanwhile March lurks just around the corner.

Winter draws to a close with reluctance. Working together, wind and snow create a winter wonderland. Here and there, drifts curl their lips in delicate curves or just pile up in dramatic depths. After skies clear once more, over a foot of white crystals blanket the ground

Over the fluffy whiteness, a weasel has moved erratically in search of prey, sometimes diving into the snow, leaving tunnels two or more feet in length. Tracking the little predator a short distance, a man observed where it leaped nimbly from the surface of the snow to the base of a small tree then down again, its hind feet registering almost perfectly in the prints made by its forefeet as it continued on its way.

Perched in a tree at woods-edge, a red-tailed hawk carefully scans the scalloped surface of the adjacent field while three crows watch it from another nearby tree. It is difficult to successfully hunt while being watched or perhaps harried by black-coated scoundrels who seem to have nothing better to do at the time.

In succeeding days, the sun shines, the air warms and the snow melts and settles. Deer move about in search of browse and some have nipped the tender ends of sumac branches attempting to appease their hunger. From one of those branches a chickadee gives the short, sweet, gurgling call that is associated with the approach of Spring. Clinging upside down from a branch it forages and calls softly several more times. A man thinks that chickadees are the best year around bird to know. They seem to always be around and are so friendly and cheerful.

Along the edge of the little ravine, numerous woodchuck burrows are now connected by brown trails in the snow as muddy woodchuck feet patter back and forth between them. A brown trail also leads to one of several small pools of water in the ravine where they have quenched their thirst.

Between the ravine and the nearby vineyard, along the headland, star-shaped tracks meander across the snow. An opossum has been out and about.

House finches sing their spring songs from the tops of trees and tall shrubs, their melodious voices carrying far in the mellow morning air. Soon the females will be carrying material to their nesting sites.

Two cardinals verbally duel as the approaching season compels them to start seriously thinking of staking out their territories. Their various calls ring out across the land adding their sweetness to those of the finches.

A sparrow hawk sits on a power line along a roadway, its keen eyes scanning the ground below. It has returned north early so as to stake out its territory for hunting as well as nesting. When the females arrive sometime later, everything will be ready.

Blue jays call, a mockingbird sits quietly in the top of a bush surveying the countryside and bluebirds warble softly and sweetly nearby.

Overhead, a large wedge of geese is heading north, trailed by a smaller contingent. Soon the sky will be filled with their northward flights, their joyous honking heralding the return of Mistress Spring. With the flight of the geese, the magic of the moment is now complete.

Along streamlets and other watercourses, there is a green freshness. Blades of new tender grasses, nourished by moisture and ever increasing warmth, peek out from their winter beds. Soon they will have grown tall enough to furnish food for deer and other herbivores.

A countryman discovered one day that there are other catkin "kittens" than the familiar pussy willows. While scanning the upper branches of a grove of quaking aspens or poplars with the binoculars, he discovered large grey "kittens" peeking out of their buds all along the twig ends and smaller branchlets. Again he was reminded how much there is to discover and learn in the natural world.

Woodchucks are also busily digging out the entrances of their burrows and perhaps doing a little early spring cleaning at the same time.

Now and then a long, wavering line of geese can be seen flying from lake to lake or from one end to the other of one. Their musical honks float down as they pass overhead. March is moving month, not only for geese but for ducks and occasional flocks of whistling swans, as well as some of the early migrant songbirds. In a few more weeks avian activity will increase tremendously. For now, the geese are congregating and moving about restlessly, awaiting the change of the seasons.

March is unpredictable. It may roar in and roar back out, or it may arrive and leave in other less dramatic ways. But there is no doubt that it is kite-flying month. It is also maple syrup-making month.

Syrup making depends on the temperature. The season may start in late February and again it may not start until March has well begun. It all hinges on cold nights and warm, balmy days. When the mercury drops down into the ‘teens or lower twenties at night and rises with the sun into the thirties and lower forties during the day, that’s when the sap rises and flows profusely from the syrup-makers’ taps.

A countryman remembers the years past when he looked forward to sap season with the sweet amber syrup and delicious maple sugar that was the certain results of tapping the trees and the gathering of sap; then there was the boiling away of the water until the syrup was just the right consistency to strain into its containers. To a man's way of thinking, the best syrup is made from smaller boilings. Much of the delicate yet rich maple flavor is lost when syrup is made in large quantities. It stands to reason that the quicker the sap is transformed into the final product, the less chance there is for deterioration and the resulting change in flavor. First run sap makes the best syrup by all accounts.

There is nothing like freshly made syrup on pancakes or hot biscuits. When a person sits down to the table with this before him, all the labor involved has been well worth the effort.

Man is not the only tapper of trees. Squirrels like the sweet sap as well, and red squirrels in particular exhibit a peculiar fondness for the sweet substance. They move about the maple trees and cut small notches on the undersides of small branches with their sharp incisors. Later, they can be observed hanging upside down beneath a limb, their little pink tongues rapidly lapping the flowing sap. When the freezes return, sapsicles hang from all their myriad taps.

The wind roars through the tree tops, it sleets, it rains, then it snows again. Before long, the sun is shining and warms the earth. Buds slowly open, sap flows, and a cardinal practices his song. It is March.

Ponderously, the wheel of the seasons slowly revolves, unmindful of the idiosyncrasy of Old Man Winter. The silver maple that the robin carols from each morning and evening when the weather warms knows that it is time to arouse from its dormancy. Already, huge crimson buds blush along each branch. Each sunny, warm day they slowly unfold a little more, yet are careful not to expose too much tender tissue to the still harsh elements that often sweep suddenly across the land. They will remain cautious until the fullness of spring bids them unfurl their blossoms.

The catkins on the female poplars are also expanding a little more each day, their soft furry "kittens" maturing in the warming sunshine.

Throughout the countryside, buds are swelling, be they flower buds or leaf buds on the trees and shrubs. Some are more reluctant than others to expose themselves so early in the season.

The dry, rustling leaves on the white oaks and the beeches continue their conversations with the breezes and errant winds, unmindful that in a few more weeks they will be displaced by the expanding buds and new leaves. For now, they are content to cling to their natal branches and continue their conversations in subdued tones.

Along a road-edge, several juncos scratch about and twitter softly amongst themselves. They are dressed in a subdued manner, the males being slate gray above and white beneath while the females are grayish-brown above. White outer tail feathers are exposed when they fly up into the nearby bushes, aiding in their identification.

A great blue heron has arrived from the southland and flies steadily along, its neck retracted into its shoulders while its long legs extend out beyond its short tail. Wind gusts buffet it as it makes its way westward. It veers a little to the left and then a little to the right as the capricious currents cause it to deviate from its course. However, the deviations are only minor and are not sufficient to deter the heron from ultimately reaching its destination.

Blue jays bugle, their crests raised in animation, their excited calls attracting the attention of many of their neighbors who look around to see what all the commotion is about. Actually, they are busily engaged in renewing their territories and their conjugal relationships.

Buoyed up by balmy breezes, a pair of turkey vultures sail serenely against an azure sky. They float effortless, their primary wing feathers curled upward at the tips as they tilt and turn utilizing every current of air to their advantage. A man is envious as he watches them masterfully sail and soar against the cerulean sky.

The sun slowly slides toward the ridges and shadows lengthen. A song sparrow sings its vespers sweetly. Chickadees and a titmouse call from the woodlot while a nuthatch laughs to itself as it clambers up and down and over the branches of a hickory. A woodpecker dips and rises in flight as it heads for its home for the night.

A few scattered snowflakes drift slowly earthward as all creation waits patiently for the winter to come to its senses and retreat to the realm of the snowy owl and the great white bear.

# SPRING

"April Showers bring May flowers" the old saying goes, but sometimes those showers are of the white, crystalline sort that seem to come and go on a rather regular basis. However, their stay is limited by warm, southerly breezes and the rays of the bright spring sun.

Rain showers, interspersed by the sun's warmth, firm up the earth that so recently has been released from the icy grip of winter.

Robins greet the dawns and end the days now with their lively carols and chirrups. The males are often seen on the lawns sparring, often fluttering several feet above the ground, beak to beak in their annual contests for territory.

Other species will soon be performing similar rites of conquest and all will become vulnerable to predators while so intensely preoccupied.

Mourning doves coo plaintively to one another while crows court in the woodlots. Killdeers fly over the fields and pasturelands uttering their piteous cries of "kill-dee, kill-dee." When they alight on the ground, they run vigorously along for several feet as if to dissipate their momentum.

Despite the irregular periods of cold that sweep over the land, each mild, sun-filled day declares that better times are coming. On such days, a fly or two may be seen flying about indolently, its somnolent buzzing a familiar summer sound. Not fully recuperated from their winter dormancy, they spend much of their time basking in the sunshine of a warm, peaceful afternoon. On the other hand, small brown ants run briskly about, attending to their affairs.

As a man walks past a favored blackberry patch, a song sparrow sings from his perch atop a tall, thorny cane. All about the countryside, each brushpile, each hedgerow and briar patch has its song sparrow, each one busily proclaiming its right of possession of that particular area.

The sparrow atop the cane continues to sing, even as a man watches close by. Later in the year, while a man is picking the ripened fruit, somewhere, in the thorny depths of the patch, a small, grass-lined nest will hold another generation of sparrows, each one clamoring for food.

Shadows lengthen perceptibly as the sun sinks lower and lower with each passing minute. Two geese fly low overhead, in a southerly direction, honking in an amiable manner to one another, one voice an octave lower than the other, providing a pleasant duet.

A cardinal calls lustily nearby, reluctant to go to his rest, while robins give their benediction to the closing day.

Dusk deepens about him as a man listens to the voices that fall upon his listening ear. Suddenly he is aware of other, more distant voices. Somewhere, from perhaps a pond or from some marshland, come the voices of springtime. The Hylas or peepers have finally aroused from their torpid slumber. Each springtime, a man waits for the medley from the marsh. The first calls may be hesitant with even a slight questioning note to them, as if they realized that they may be silenced several times by recurrent cold weather.

A man knows, however, that once awakened, that he will hear them more frequently as the weeks progress. Their voices pulsate through the evening air and a man is content in the knowledge that all of creation is awakening to newness of life and activity.

Mild spring evenings and spring peepers go together and are two of the things a countryman relishes after a long, cold winter. Although the evening temperature may slowly slip towards the thirty-two degree mark, the little tree frogs call lustily from their wet world. An old country saying states that the peepers have to freeze up three times before it is truly spring. Over the years a man has seen seasons when cold snaps have hushed their voices more times than three, yet as soon as warmth returns so do their activities, often with even more vigor than before.

The temperature continues its decline and flakes of snow appear in the crisp evening air as moisture-laden clouds move in. The shrill voices of the hylas diminish and trail off until only one or two continue until they too, fall silent as the snowflakes thicken about them.

During the night, the snowfall intensifies and continues on into the succeeding day, decreasing and increasing in intensity in proportion to the amount of moisture present in the eastward-moving clouds.

Crows stand out distinctly against the white background while other birds, even the crimson cardinals, are merely dark silhouettes amongst the falling flakes.

Song and house sparrows hop beneath the shrubbery along the patio edge. A man throws out some millet seed and soon they are feeding hungrily. Later on when the others have eaten their fill, a demure junco comes in and eats quietly, always alert for a sign of danger. The robins sit dejectedly in bushes and shrubs, their lawns and fields now blanketed with snow.

The moisture-depleted clouds move away and the sun peeks through the overcast sky. As its warm rays reach earth once more, the snow begins to melt and open patches appear on lawn and field and pasture. It does not take the birds long to discover these open areas and to begin their renewed search for food to fill their now empty stomachs. Flocks of starlings and cowbirds converge on many of these opening areas and the ground darkens with their numbers.

All nature becomes more active as the air warms once more to a more comfortable degree.

A Cooper's hawk sails overhead on winter-worn wings. A primary flight feather or two is missing from one wing but that does not seem to impede the bird's flight. The missing feathers and an overall ruffled appearance gives to the hawk a rather ragged appearance as well. This no doubt is due to its many swift forays through the brushlots and among the tree trunks of the woodlands in pursuit of prey, as is the custom of these accipiter hawks. For a brief time this day, it soars and flaps its way over the countryside, the air whispering over the surfaces of its outstretched wings.

A crow has gathered several long twigs in its beak and flies to another nearby tree. Placing the twigs safely beneath a sturdy foot, it caws several times. It looks about and perhaps listens for a few moments. In the distance, another crow calls once, twice. Perhaps it is the first crow's mate. Retrieving the twigs deftly with its beak, it flies across the field to another woodlot where it is lost to sight.

The sun shines between fluffy white clouds moving serenely across the sky. A cardinal calls and is answered almost immediately by a rival. The nest-building season has begun. For some at least.

The rains and showers of April revitalize the earth and all growing things. Water seems to flow almost everywhere and where it does not flow it relaxes into pools and puddles and creates temporary marsh-like areas.

The rains fill wells and rejuvenate springs that had been at a low ebb during the cold months. Seasonal springs, those that are active only during wet periods, now well up in unexpected places. As they bubble to the surface thy give the illusion of boiling and seem to take on a life all their own. As they escape from the ground, they bring with them grains of sand and minute debris that forms small cones beneath the surface of the water. Like miniature volcanoes that have substituted water for lave and cinders, they continually build and change until their source retreats or dries up as the season progresses.

Water seeks its own level and all its numerous sources flow and merge and intermingle, streamlets become small brooks which strengthen into larger creeks. These, in turn, become tributaries to other creeks which become rivers. These also strengthen in size and current and flow ever onward to eventually disappear into the waves of some far off, distant sea.

Too often we take for granted the endless cycles of life about us. Only when there is a lack of moisture or there is too much do we come to a realization of how much our lives are intertwined with the forces of nature.

The evaporation of moisture from the great bodies of water, carried by the clouds thousands of miles inland to be dropped as rain or snow, is only a part of the great cycle that helps sustain all life.

The rains and snows that fall on the highlands first trickle then tumble and finally rush and roar headlong to the valleys below where they join forces with the moisture that has also fallen there.

Watercourses are the nourishers of many forms of life, animate and inanimate. Grasses, flowering plants and small shrubs partake of the water and in turn send their roots deep into the ground holding it in place resisting the erosive force of the stream or river. Trees provide shade for the lesser plants and their branches and trunks provide homes for many species of bird and animal life.

The leaves expel moisture into the air as well as oxygen, thus completing the cycle. This is why a wooded area feels so cool; it is one of nature's air conditioning systems.

Our world is controlled by many cycles, some small while others are global in extent and influence, yet each one contributes to the whole.

The hand of the Creator can be seen sustaining these endless and complicated movements that nourish and strengthen all the life-forms that inhabit the earth.

As one observes the many facets of nature and contemplates all the actions and reactions involved, there is only one conclusion that can be drawn and that is that there is a Master Worker involved in every minute detail of life as we know it.

Warm air pushes up from the south-lands; rain clouds move eastward allowing the sun to shine once more upon the land and several days of unseasonably warm weather ensue.

Mistress Spring has been waiting in the wings for just such a time as this and now she steps out upon a verdant stage to direct the production.

Pussywillow catkins have enlarged and matured into elongated forms that glow golden with pollen, attracting numerous small insects including a hornet or two.

Poplar catkins dangle like so many long, fuzzy caterpillars that wave at an oblique angle in the warming breeze, resembling pennants until the breeze dissipates and they assume their former appearance. They also are laden with pollen which attracts insects which, in turn, provide food for birds such as phoebes and tree swallows. Already a phoebe can be heard calling about a nearby farmyard. In a few weeks its mate will have arrived and no doubt will be selecting a sheltered spot for her nest beneath the eaves of one of the outbuildings. Tree swallows will arrive most any day now, followed by the rest of the swallow tribe.

A queen bumblebee, newly-emerged from hibernation, buzzes noisily about while sleepily seeking some form of sustenance. In the days to come she will become more invigorated and will begin earnestly to search out a place to raise her family. For now, remaining airborne consumes much of her energy.

Facing the morning sun, a meadowlark calls from its perch atop a tall spruce tree. Its breast glows golden in the sunlight and is adorned with an ebon V-shaped marking. A meadowlark, a man thinks, is the embodiment of open spaces and spacious skies. Its cheery calls remind a man of years long ago when he first heard the golden-breasted lark singing from atop a fencepost bordering an old hay meadow.

Where else but from a field should one hear the first field sparrow? Soon they will be singing their sweet but slightly sad song from all over the nearby vineyards.

The sun seems to almost pull reluctant leaves from their protective sheaths. Leaflets peek out along a cherry branch and turn the large willow trees from a golden flush to an effusion of pale green. A few bushes growing along a southern exposure are already clothed with tender leaves.

A fluttering movement in one of these bushes attracts a man's attention. He watches for several moments and then a mite of a bird appears mometarily then disappears again into the greenery. It was olive green above and creamy below with wing bars. It re-appeared again briefly and was observed more closely this time. It was a female ruby-crowned kinglet as there was no trace of ruby on its diminutive crown. The male, although able to nearly conceal his scarlet feathers from view, cannot resist flashing them from time to time and, during courtship or times of aggression a small crimson crest is erected to proclaim his authority.

A woodchuck sits atop the mound of dirt at the entrance to his burrow and watches as a wild turkey scratches industriously at the edge of the field. From time to time, the woodchuck closes his eyes as his russet coat absorbs the warm rays of the afternoon sun. He would like to really fall asleep in the sun's all enveloping warmth, but his instincts warn him that it would be dangerous for him to do so. Instead, he just dozes coming alert every few moments to observe his surroundings. To be sure, it is a fine spring day.

Around the last of April in any given year a countryman will notice some radical changes and these changes become more rapid and increasingly noticeable as the season progresses into May.

The earth is now carpeted with soul-soothing green. Wild mustard flower buds nestle in the sheltering cups of new leaves and each day leaflets unfurl a little more along the bramble canes. Many shrubs and small trees are shrouded in various tints of green.

Burdock leaves rise from their winter beds and slowly unfold. Soon they will grow into sturdy albeit noxious plants. Catnip plants emerge from soil where seeds were strewn the previous year by the parent stock. Soon, they too will rise on square stems characteristic of the mints.

White violets now brighten the brown carpet of the woodland floor and slender spikes of Lily-Of-The-Valley rise to view from the woodland's edge.

A brown thrasher calls from the top of a sapling. He is dressed in a russet coat and creamy vest adorned with heavy streaks of the same color as his coat. There are two creamy bars on the shoulder of each wing and he views his fresh new world through bright yellow eyes.

The thrasher is a mimic thrush as are his cousins the catbird and the mockingbird and some of his repertoire resembles some of those of the catbird. All three are strong-legged and sport noticeable long tails.

Depending on the weather and the abundance of insects, the tree swallows arrive by late April or early May. They gracefully swoop and soar uttering cheery notes that sound much like "Pretty good! Pretty good!" as if they find everything much to their liking.

Originally nesting in hollow trees, they find bird houses very suitable since there are few available tree cavities today. Houses put up for bluebirds are quickly taken over if they are placed very high above the ground.

Beneath bushes and along hedgerows, white-throated sparrows busily scratch and often stop long enough to sing their sweet song of admonition, "sow wheat, Peverly, Peverly," In New England it is often translated as "O sweet Canada, Canada, Canada!" At any rate, they are a joy to hear.

A small brown bird with white outer tail feathers flushes from the ground near a brushlot. A vesper sparrow had been taking a bath in a rain puddle and now fluffs its feathers and carefully preens them in to place as it sits in a small tree. This is a bird of field, meadow and road-side and wears a grayish-brown coat streaked much like a pale song sparrow. Unlike the song sparrow, it has a light eye-ring and small patches of chestnut on each shoulder.

Chipping sparrows trill from their perches around the door-yard and along the pasture edges. They are tame little fellows with chestnut caps and a white line over each eye.

Evening grosbeaks, those big tan and yellow birds with the large olive-colored beaks, seem reluctant to go home to the northern coniferous forests. As long as there are sunflower seeds for them to gobble at the feeders they may stay until maternal instincts advise the females to urge the males to join them in their journey north.

A countryman enjoys this time of the year when nearly every day brings one or more different birds north for the summer. This is a fine time for the beginning birder to take field guide and glasses in hand and go afield to see what new species he or she can discover, and enjoy the many other pleasures the outdoors has to offer.

There comes a morning in May when all the world seems to proclaim that Spring really has arrived. A few cumulus clouds, scattered across the sky, greet the morning sun. A cabbage butterfly flutters here and there seeking out blossoms.

Flickers call vigorously; the males bob and bow before the females hoping to create the right impression and in the process win a mate. They hitch up and around a tree trunk seeming to play hide and seek. On occasion, a second female arrives making matters more interesting if not a bit confusing for the males. No doubt that is all a part of their mating game.

While flickers occupy the mid and upper levels, the undergrowth seems to be alive with catbirds this particular morning. One calls from a witch hazel tree while two play tag in and out of the bushes and a third stops its calling to join in on the activity. Their calls are interspersed by plaintive mews, for which they are aptly named.

A cardinal, crimson crest raised perkily, glows in the morning sun while he proclaims his territorial rights from a branch half-way up a tall shag-bark hickory. Although his calls have a purpose, yet to a listening ear they are melodious and ring out cheerily on the morning air.

Two blue jays fly about the countryside uttering muted, yet bell-like notes that are in sharp contrast to their normal, harsh calls. Where one goes, the other follows as though each was connected to the other by some invisible tether. Why of course, it is springtime!

A red-bellied woodpecker alights on a willow tree and commences a thorough inspection of its branches. Its churring call is added to all the other voices being heard across the land. The red feathers on his head and nape blaze briefly in the sunlight as it turns aside to inspect a possible nesting site in one of the dead limbs nearby.

Sweet song bubbles from the throat of a little house wren as it busies itself in the leaf-litter in the hedgerow. Later on, it can be heard chattering away in the sumacs at the edge of the field.

Out by the pasture another new arrival can be seen perched on the electric fence. Dressed in a gray jacket and a white waistcoat, the kingbird surveys his kingdom while waiting for an insect meal to fly by.

Up on the nearby powerline a meadowlark and a red-winged black-bird sing an unbalanced duet that adds variety to the morning symphony.

Over the pasture and adjacent meadowlands, fork-tailed barn swallows skim and dip, their musical twitterings seeming to hang in the morning air. Now and then they are joined by several tree swallows, and together they scour the air of insects. Soon they will be busy clearing mosquitoes and other noxious insects from the environment. Their voices add a touch of summer yet to come, yet proclaim that spring has indeed arrived.

As a countryman watches and listens to these myriad sights and sounds, yet another voice is heard, different from all the others. From the depths of the woods comes the dulcet, bell-like notes of the wood thrush. There is an ethereal quality to this voice, a quality that tends to lift one's spirits above the mundane to a higher, purer, spiritual plateau. A woods, a man thinks, is not complete without the melodious, silver-toned voice of the wood thrush.

The morning air warms rapidly, the humidity already building to noticeable levels. There could be a thunderstorm before the day is over, but at the present it is one of those wonderful spring mornings.

An osprey soars high overhead. It is recognized by the light underwings with a dark patch at the "wrist" of each wing. The wings are held with a slight angle or crook to them. There also is much white about the head but two swatches of black through the eyes create a distinguishing pattern that identifies the bird as an osprey and not an eagle for which it is often confused. Although it is only three miles from the lake, it is a bit unusual to see this fish hawk soaring over fields and woodlots.

A female marsh hawk dips and soars low over a fallow field in search of meadow voles while at a higher elevation the local red-tail soars on warm up-drafts from the warming land. Its mate is busy incubating her eggs in a large stick nest somewhere in a nearby woodlot.

Dogtooth Violets nod their yellow heads in the morning breeze. Chokecherry and black cherry blossoms bloom profusely, their thick, white blossoms veiling the branches as tiny new, pale green leaves slowly unfurl unobtrusively behind and beneath them. Their perfume pervades the air while bees and other insects are attracted to them.

The blossoming fruit trees attract the insects which, in turn, attract the migrating warblers. A Nashville warbler peers into a row of blossoms while a yellow warbler flutters about from limb to limb, stopping momentarily to sing its sweet song. Near the top of the trees a black-throated blue warbler busily picks insects and larvae from the twigs and amongst the flowers. A male redstart flashes the red-orange feathers in its tail and flutters its wings as it too feeds hungrily after a long night's flight.

A northern oriole forages among the branches, cheerily calling to all around him. His bright orange and black suit contrasts against the white of bloom. White-crowned sparrows practice their spring calls.

Bumblebees hum sonorously as they gather nectar and distribute pollen from tree to tree, ensuring an abundant crop of fruit. Honeybees also contribute their very important part in the pollination process.

By mid-afternoon, dark clouds have gathered along the western and northern horizons. The air hangs heavy with humidity and thunder mutters in the distance. A hawk rides the thermals that emanate from the now heated land.

Lightning streaks down to a distant ridge and a few moments later a heavy rumble of thunder is heard.

In a rhododendron, a robin covers the eggs in her nest as scattered raindrops splatter the nearby leaves and splash on to her back as well. She settles further into the nest, prepared to ride out the storm.

The northern sky darkens perceptibly, outlined and accentuated by aquamarine and turquoise shades mixed with gray. Thin grayish-white streamers of mist and light rain drift eastward and soon develop into a definite curtain of rain. Thunder grumbles and rumbles ominously as the storm strengthens and rapidly approaches.

Brilliant flashes of lightning are more frequent now, quickly followed by sharp reports of thunder. The wind has increased and trees bend before its force while the rain is now falling steadily and is blown into oblique sheets by the wind. The falling water sweeps across the land, filling ditches and raising the levels of all the small streams.

A countryman retreats to the shelter of his home and listens to the thunder and the driving rain against the windowpanes as a spring storm makes its way eastward, refreshing the land.

The land is full of activity by the time mid-spring arrives.

On warm sunny days when a breeze blows softly from the south, numerous warblers may be seen migrating northward. They fly at night and feed and rest during the day, perhaps gaining a little in their movements.

Wearing a necklace of black streaks across its bright yellow breast, a Canada warbler forages in a sapling veiled in new light green leaves. Male yellowthroats call "witchity, witchity, witch" and peek out from the bushes like little masked bandits, their black face masks contrasting sharply with their olive green backs and yellow throats. Yellow warbler males flash back and forth between the trees and bushes as they attempt to settle territorial disputes. Cowbird females sit silently nearby, waiting patiently for nests to be built so that they may lay their eggs and let other birds assume their parental responsibilities.

A small flycatcher energetically pursues insects flying past its perch. It is one of several small, look-alike Empidonax flycatchers that can only be identified by their voices. This one is not saying anything. It is too preoccupied with catching breakfast.

A pair of catbirds industriously carry nesting materials to the interior of a dense bramble bush which will provide protection for them.

The nest in the rhododendron now contains three baby robins who nearly fill the nest and rest their chins on the rim as they wait for a parent to arrive with food. In yet another shrub, a pair of robins are just putting the finishing touches on their nest.

Starlings and grackles are also feeding nestlings while other birds are still incubating and yet others have yet to start building nests.

Two indigo buntings search a brushy woods-edge for a meal. One pauses for a minute or two to re-arrange and preen its feathers. These buntings are mostly all blue but of a deeper shade than that of the bluebird. Indeed, they are an all blue bird, belonging to the large finch family. Since these are both males and seem to be companionable, perhaps they are migrating further north rather than settling down in this area.

A male scarlet tanager flies overhead in a purposeful manner, its brilliant red body accentuated by ebon wings and tail glowing in the morning sun. Once common, their numbers are diminishing due, in part, to habitat destruction, both here where they nest and in their tropical wintering grounds. This is also true for many other songbird species.

Maple leaves hang like miniature umbrellas while those of the quaking aspen rustle together in the breeze creating a sound much like a conversation that is not only animated but also of a pleasant nature.

Hovering over a field on rapidly beating wings, a rough-legged hawk scans the ground below for prey to sustain it on its northward journey.

While all this activity is going on, little blue violets quietly carpet the ground in places and yellow mustard boldly marches across the countryside, hand in hand with the dandelions, some of which are already yielding seed-laden parachutes to the wind.

Everywhere a countryman looks, he can see the hand and mind of the Creator at work, renewing, replenishing and sustaining the world around him. How grateful we should be for that constant, unwavering Presence.

While the cherry blossoms have fallen like snowflakes, covering the ground beneath the trees, apple and lilac blossoms fill the air with their fragrance.

A black-billed cuckoo gives its distinctive call as it seeks the newly-hatched tent caterpillars that so destructively defoliate trees early each spring. The cuckoo clan is one of the few bird families that will eat fuzzy caterpillars and they especially like the tent variety.

Pausing on its northward journey, a male purple finch sings its rich melody. A countryman remembers their sweet voices when he lived in their home range in New England. Their cousins, the house finches, have replaced them to a great extent in the middle Atlantic states, taking over their territory and their food supply. Then too, the purple finch is a more northern species and prefers a higher elevation.,

The bubbling voice of a bobolink comes from the meadow-lands, so bursting with energy and so full of life that it is bound to lift one's spirits.

The dandelions are rapidly going to seed, but their golden blooms are being replaced by the shiny yellow blossoms of the buttercups.

The nest in the rhododendron is empty and tilts to one side now that the young robins have left. They can fly quite well now but for only short distances. Each day their wing muscles will strengthen and before long they will be on their own. For now, they are keeping their parents busy providing nourishment and watching out for enemies.

Strawberry blossoms parade over the ground, their yellow centers brightening the white-petaled cups of each bloom. In another month the scent of ripening strawberries will be wafted on every breeze.

A female yellow warbler busily builds a cozy cradle in a low bush, half hidden by new green leaves while her mate sings and forages close by.

May apples or mandrakes grow profusely in the shelter of a woods-edge. Each plant has two large, umbrella-like leaves. From the junction of these leaves grows a short stem on the end of which is a large white blossom which will later develop into a yellow-green fruit.

Greenish-yellow spires of new growth appear like candles at the tip ends of the white pine branches, while the branches of the spruces and firs are outlined with similar new growth but in a different shape. Atop each conifer a new shoot has developed that will eventually add to the height of the tree. Soft, small new cones are also developing on some of the spruces.

Everywhere one looks there is much activity as May concedes to June its procreativity. New life, animate and inanimate, is seen everywhere one looks. There is all stages of development from embryo to maturity, from the very simple to the extremely complex. And there will be more to come as the season progresses. This is the time of year when the hand of the Creator can be seen at work in all its complexity. Re-birth, new birth, regeneration – all is an indication of a Master Mind at work.

When a countryman thinks of June, several things come to mind. There are June berries, the small but tasty fruit of the shadblow or serviceberry tree that the birds seem to always get to first before a man can get more than a taste.

Then there are June bugs, those big, clumsy, blundering beetles that buzz about the screen door or window screen on a warm June night and are relished by the leathery-winged bats that catch them in mid-air, shuck off the wing-covers and devour the beetle before the discards reach the ground.

However, before the berries ripen and the beetles emerge from their natal quarters beneath the ground, a man takes to the fields, bag in hand, to pick tender milkweed shoots for a mess of early greens.

May produces its dandelion greens, but June provides the milkweed, a mild and tasty green that goes well with mashed potatoes. Later on, it will produce pink blossoms that will scent sweetly the air and be welcomed by both butterfly and bee.

The erstwhile brown fields and bare-limbed trees are now clothed with luxuriant verdure, the many shades of green blending into one harmonious whole.

The grasses grow thick and lush and some already are striving to reach maturity while others are content to attain that stage more slowly.

The migrant birds, for the most part, are no longer in the area and the residents are busy with their families or are nest building. The barn swallows are pairing up and pairs can be seen perched together on utility wires, twittering softly and meaningfully to one another. Soon they will be plastering their mud houses beneath eaves and against the rafters of outbuildings.

Phlox, in its varied hues ranging from white through pink to deep rose, blooms profusely in patches of color, its scent pervading the air.

The willow bush that held the yellow warbler's nest now holds a different surprise. Apparently a female cowbird had discovered it and had laid her egg in it. Not to be put upon by the lazy cowbird, the warbler built another nest over the first. The cowbird apparently laid another egg as a third nest was built giving the whole affair a three-tiered aspect. It remains to be seen how it all will turn out.

Passing an overgrown watercourse, a man flushes a bright blue bunting. It is not frightened, but rather busies itself by picking beetles from the branches of a nearby elm sapling. Moving closer to observe the male, a man disturbs the female who had been taking her bath in the ditch. She sits a few moments to re-arrange her wet feathers and give them a good shake and preens them into place. While she is performing her toilet, a man shifts his gaze to where the male was and he has disappeared. Looking back towards the female, he discovers that, she also, has taken her leave. In the distance, a cheery bunting voice is heard, and whether the voice belongs to the recently observed male or to yet another, a man relishes the sound of late spring along a woods-edge.

Moving through the waist high grasses near a hedgerow, a man now disturbs a napping doe. She bounds away, her white flag of a tail waving jauntily as she ducks through an opening in the hedge.

A countryman looks to see if a fawn lies hidden nearby, but sees nothing but the now-empty form where she had lain. Looking about further, three more older beds are seen, indicating that she found the area much to her liking.

A man reflects that June is many things. June is young robins on a lawn, begging to be fed. It is also oriole songs in the morning and thrush songs in the evening. June is all of this and much more.

It seems that June could aptly be called "Woodchuck Month", for everywhere one looks there seems to be a woodchuck in sight, if only briefly.

One runs across a vineyard road and disappears in the tall grass on the other side. Another stops chewing on tender clover shoots, along a vineyard headland, to sit up and watch a man meander by. Passing by a woods-edge, the shrill whistle of yet another one comes from the protective cover of its burrow.

A mother woodchuck discovered an old trash pile covered with brush to make her nursery. The pile is overgrown with wild grape vines and Virginia Creeper vines, creating what might be called a woodchuck castle. Several well-traveled paths radiate in different directions from this country castle.

One hazy, humid afternoon, a countryman was homeward bound when he noticed movement near the top of the castle. Through his binoculars he watched the family for some time. The mother watched from the top of the pile while at least two half-grown youngsters explored or sunbathed nearby.

One youngster nibbled tentatively on a creeper leaf, but it wasn't to its liking so it tried a grape leaf which went down a little better. Then it tried a piece of Styrofoam which it soon lost interest in.

Disappearing into one opening, it emerged immediately from another one close by. Then it rolled to its back and, grasping one hind foot with both of its front feet, it chewed what apparently was an itch, then it was up and exploring again. Meanwhile, the mother and the other sibling had disappeared so I continued on my way.

Another day, I was about to cross a deep ditch when a 'chuck ran from nearby in a tremendous hurry. When it reached the edge of the ditch it tumbled head over heels to the bottom, lying there for a moment on its back, totally confused by its tumble as well as by my presence. Then it was back on its feet, dashing right past one of its entrances, raced around a bramble bush and finally dove into yet another hole. To that woodchuck, it all was a terrifying experience, but to a countryman the whole episode was hilarious. It just depends on what happens to whom.

Woodchucks seem to be constantly cleaning house or renovating their quarters. Every day or so, fresh dirt and stones appear at the mouth of their burrows, except for the secret back doors that they usually have. These are just round holes in the ground with no tell-tale dirt in sight, the stones and soil all carried back and out another entrance.

These are emergency entrances to be used if a woodchuck is caught unawares too far from its usually used one. This is how a ‘chuck can be seen feeding or sitting up and then it suddenly disappears – vanishing as if into thin air.

For all its destructive activities in the garden and the problems it causes the farmer, the woodchuck is a very interesting animal to observe. Their front feet are used as “hands”, grasping the stem of a plant while it eats the leaves, and as often as not the stem as well. These front feet are used for grooming its coarse coat as well as digging, and aids in climbing trees. They do not climb like their cousins the squirrels, but they will climb to low branches when the urge comes upon them to do so. A man has observed them do this on several occasions.

Yes, woodchucks are interesting, and can be very entertaining.

# SUMMER

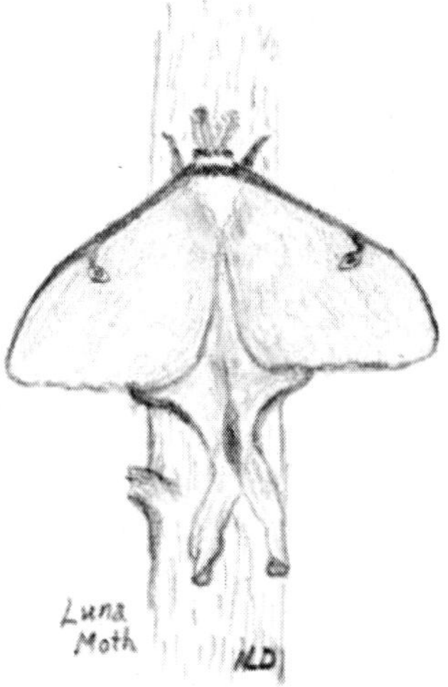


The sun has once again reached its pivotal point in the heavens, declaring to one and all that summer has finally returned.

The days have stretched to their limits and the long hours of daylight have produced tremendous growth in all vegetation. The grasses, in a few short weeks, have grown to the point where the scent of newly-cut hay permeates the air. Bales of hay repose on the breast of many a field.

In those fields that have not been cut, bobolinks are busily tending their nestlings, their bubbly voices now silent as they go about their duties.

Marsh hawks dip and sail over fallow fields which hide their humble homes hidden amongst the grasses.

Red-eyed vireos sing repetitiously from the woodlands, their earnest voices earning them the name of "Preacher Bird".

A sneezy call of "fitz-bew" comes from a brushy field. A willow flycatcher, one of the Empidonax flycatchers, is in the area, identifying itself by its distinctive call. Flycatchers concentrate on small, flying insects as their source of nourishment.

Lurking in the leafy shadows, a yellow-billed cuckoo calls mysteriously, "ka-ka-ka-kowlp, kowlp, kowlp, kowlp", the syllables slowing towards the end. This cousin of the black-billed cuckoo also seeks out hairy caterpillars on which to dine and feed its young.

For several weeks, the wild roses have bloomed and shared their beauty and delicately sweet perfume with all who passed by. Now, as the days of spring merge with those of summer, their flowers will also give way to fruition.

Here and there, an occasional yellow or black swallowtail butterfly will be seen floating and dipping over the countryside, stopping at one of the numerous blossoms for a sip of nectar. A lone monarch checks out the progressive growth of the milkweeds. Sad to say, butterflies are not anywhere near as abundant as they used to be. Insecticides are to blame as well as a loss of environment.

The titmice and chickadees quietly go about their family duties. On a rare occasion a countryman will hear a chickadee call "knee-deep" several times from a leafy tangle, and a man will chuckle to himself as he wonders what the little fellow could possibly be knee deep in now that the snow has long since disappeared. Perhaps it is its way of expressing joy over all the bounties of the season. Perhaps he is knee deep in just plain contentment. Now that is something worth singing about!

Strawberries ripen and shortcake is the order of the day. There is nothing that tastes any better than freshly baked baking powder biscuits smothered with wild strawberries and topped off with milk. A countryman can not think of a better way to close one season and begin another!

As another day draws to a close, a robin starts its benediction as the long shadows creep over the land. This long summer day began with robin-song shortly after four in the morning and now, after nine in the evening, the robins' song is heard once more.

The first few fireflies wink in the gathering darkness; tree frogs call throatily to one another as the summer season begins.

The warm, humid days of June serve as a catalyst for the occupants of our native silkworm moth cocoons. The pupae rapidly develop after their long winter dormancy in their snug, insulated silken shelters. They twist and turn restlessly as cells and chemical compositions mix, change and multiply to form the large, colorful, night-flying moths of late June and early July.

Less than a year ago they were caterpillars or larvae, feeding on the leaves of various trees and shrubs such as maple, oak, walnut, wild cherry, sassafras and privet, to name a few.

From tiny eggs laid on favored leaves, the diminutive caterpillars emerge and immediately commence feeding. They will eat until their suits are filled to bursting, then they will rest a short time. Soon, their skins start to split from the head end and they wriggle and crawl their way out, emerging wearing new baggy suits, often of a different style. This process of eating and moulting their skins will occur three or four more times before they are ready to spin their cocoons. Each time they change suits, the style may change subtly or dramatically, depending upon the species involved.

The larva of a silkworm moth eats not only to sustain life as a caterpillar but also to transform from the pupal or resting stage to its final form as a moth. These large moths do not eat and their lives are brief, averaging two to three weeks. During this time males seek out the females, they mate, the females lay their eggs and their duties are over. If perchance a moth does not mate, they fly about until they weaken and die, just as those who successfully mate do.

The life cycle of moths and butterflies is called metamorphosis, which means a change or transformation from one form to another.

The lowly, crawling caterpillar changes into a pupa. The pupal stage is similar to death, yet in this case there is an element of life. The caterpillar is gone forever and that which is within the pupa case is not a moth or butterfly. Nevertheless, when the appropriate time arrives, there is a resurrection of sorts and a beautiful creature emerges, complete with wings. No longer bound to creep about, this new creature has a totally new body and when the wings have fully expanded and hardened it is able to fly away to fulfill the plan programmed within it by its Creator.

Whenever a large, gray, tan, red-brown or green moth flutters about an outside light, or clings to a door or window screen, remember its humble beginnings, its transformation, and marvel at the wonder of it all.

The seventh month is a proclaimer of several things. First of all it is the first full month of Summer, secondly it is the first month of the second half of the year.

The sun has reached its seasonal zenith and now is very slowly, yet inexorably, starting its southward swing. However, a countryman anticipates many an enjoyable, long summer evening between now and first frost.

The wild strawberries are pretty well gone but other fruit are now becoming ready to take their place. The pale green buttons of the black raspberries are tinged with a blush of rose. Some have even taken on a deeper hue, a portent of things to come. After each summer shower the yet green fruits of the blackberries swell in size a little more, aided by the hot rays of the summer sun. Within a few more weeks they will be ready for sauce, pies and to be frozen or canned to be savored as a bit of sweet summer essence later during the frigid days of Winter.

Insects are increasing in varieties as well as numbers and will continue to do so until the first killing frosts take their toll. Mosquitoes make certain outdoor activities a challenge as do the gnats and pestiferous deer flies. Beetles, moths and other tiny insects accumulate around the outdoor lights, the smaller ones creating a feast for the opportunistic toads that sit beneath the lights waiting for a snack to drop down in their direction. As twilight deepens into darkness, the phosphorescence of the fireflies twinkles all across the landscape.

A doe deer leads her twins from one succulent pasturage to another. The fawns gambol and play, innocent of the many dangers in the wide, big world that surrounds them. For now, life is made up of adventure, play and mother's milk with perhaps some tasty greens added to their diet as they mature.

Buck deer are busily feeding and growing new antlers beneath blood-enriched velvet. A forkhorn buck stops eating near a field road to watch a man's approach. Moments later, he disappears into the nearby brush, his white tail waving jauntily back and forth.

A countryman bends to pick the last few scattered wild strawberries along the edge of the field and now and then he can hear the young buck moving about in the brush. An occasional snort from the deer lets him know that he is interrupting its evening meal.

The sweet-scented wild roses of June have given place to blue-eyed grass which peers up at one from atop slim, tall stems. Some of their blooms are truly blue while others are of a magenta hue. Buttercups slowly surrender their bright blooms to the colorful orange and yellow hawkweeds, or paintbrush.

The baby yellow warblers have outgrown their three-tiered home and are now on their own.

Noisy grackles lead their young in a phalanx across lawns and up and down vineyard rows in search of grubs and beetles.

July has a special meaning for this countryman as both he and his mother were children of the seventh month.

Early summer paints a pleasing picture. It starts with a background of hills and ridges, their usually sharp outlines now softened by haze. The haze extends into the sky creating a pale blue dome with puffy cumulus clouds painted across its vastness. Freshly cut hayfields and croplands of varying hues of green provide a foreground ready to be filled with many natural things of interest. And then too, there are numerous niches waiting to be occupied.

Along the field edges, untouched by man or machine, are many forms of plant life. Field daisies, with their friendly faces, are a joy to behold while the deep, orange-yellow blossoms of the black-eyed Susans add their warmth to the scene. A countryman also notices the compact, white flower-head of yarrow and will bend to crush a few leaves so as to enjoy its pungent odor.

Cerulean blooms of chicory grace their sturdy stems as they grow by the roadside, along the wayside and across the waste places. Milkweed finds these same places a favorable environment and the fresh, new pink flower heads lend their perfume to the air. Black and red milkweed bugs amble across the broad leaves, but nimbly scurry underneath if disturbed.

Damselflies and the larger ten-spot dragonflies skim over and hover about damp meadowlands and wet areas, foraging for flies and other insect prey. Now and then one will alight on a flower head or a stem of Timothy and promptly set its wings into their resting position with several short, well coordinated muscular jerks.

The once small paper nests of the wasps have increased in size and do so on a daily basis. For the most part, these gray paper homes are hidden up amongst the leafy branches or down low to the ground in some small thicket or even in some subterranean cavity.

An orange-brown fritillary butterfly uncoils its long tongue to probe the depths of some of the deeper flowers as it satisfies its need for nourishment. It drifts about among the many blossoms, sampling the various nectars as it goes.

By now the scene is nearing completion, only a few brief touches here and there are now needed.

A male oriole, in his resplendent black and orange suit, chatters from a momentary perch on a guy-wire before flying into a nearby tree. Overhead come the rollicking notes of several goldfinches, enjoying the summer day. A yellowthroat calls mysteriously from a densely-leaved thicket.

For one final touch, one might add the sweet yet somewhat pensive voice of a field sparrow, somewhere out in the vastness of a vineyard.

A countryman does not need a calendar to tell him that July has already reached its intermediate position. For the most part, Nature is right on schedule with her seasonal activities. Of course there will be a year, now and then, when early warm Springs or late frosts may advance or set behind things a week or so. But for the most part she is a very punctual lady.

A countryman knows, for instance, that when the elderberry blossoms turn yellow and their tiny petals fall away that it is time to get his dish and go pick the first few black raspberries that have just ripened.

He knows too that the May Apples or mandrakes will now have pale green fruits whose cheeks already may have a flush of yellow indicating that they are commencing to ripen.

A man looks too for the wild morning glory or bindweed along the grape rows and is not disappointed in his search. Their white and pink trumpets beckon to the bees as they clamber up a grape post to intertwine their tendrils with those of the grape vine.

Then there are the teasels. A few short weeks ago one would never suspect that they were patiently waiting to keep their appointment with the season. Now they have grown up above the tall grasses on their stiff prickly stalks, their cone-shaped flower heads already starting to show a hint of lavender bloom that will reach maturity in the days and weeks to come.

Young gray squirrels, now nearly grown, feast on maple seeds which fell to earth back in May. When not eating, they run up and down the numerous tree trunks and play tag among the branches and then run back down to the ground to scamper across lawns or woodland floors.

By now, chickadees move about in small family groups and communicate to one another in subdued tones as the youngsters learn how to fend for themselves.

Likewise, a mother downy woodpecker tries to teach her youngster to become self sufficient, but he noisily clamors to be fed. When she ignores his petulant cries for food, he hitches slowly up a dead sumac limb giving it half-hearted pecks as he goes, causing small pieces of bark to fall to the ground below.

His adolescence is further assured when a man notices the red-orange feathers on his crown. Eventually this will be reduced to the narrow crimson strip further back on his head, indicative of full maturity.

Field corn is waist high or higher, the broad green leaves rustling as a summer breeze cavorts in and around their sturdy stalks.

Yes, a countryman knows what time of year it is when he also sees the green globes among the branches of the black walnut trees. The calendar of the countryside boldly proclaims that it is indeed half-way through July.

Summer sights and sounds are befitting the season. A rambunctious thunderstorm may flash and crackle its way across the land. The storm clouds having recently passed, the sun shines brightly on the departing curtain of rain, creating a beautiful rainbow whose broad arch across the sky lends a softening and subdued ending to the turbulence of the recent storm.

Green frogs float just beneath the surface of a farm pond and utter their plunk-plunk-plunk call as they wait for flying insects to come within the reach of their long, sticky tongues.

Barn swallows skim low over the pond to snatch up the insects that have managed to escape the frogs, while along its damp edge yellow sulphur butterflies stand with folded wings to imbibe of some of its moisture.

Bluebirds warble from an overhead wire, their familial duties now ended. Their youngsters are no longer dependent upon them and the small family groups now roam the countryside, sometimes separately, sometimes together.

Robins have raised two broods and, on occasion, a pair will go ahead with a third one if there is plenty of food available.

Relatively quiet the past two months while raising their broods, bluejays have become quite vocal of late, accusing anyone and everyone within earshot of larceny as though their own records were spotless.

Coming around a large pile of brush, a man surprises a doe as she feeds beneath the branches of a willow. Their eyes meet and she whirls, to dash into nearby cover. He continues on his way along a circuitous trail until he reaches a grove of spruces. Ever aware of his surroundings, he notices a vague shape beneath one of the conifers and stops for a better view. It is the doe, head lowered on an extended neck, peering curiously at him. As he starts forward, the spell is broken and the doe leaps out and across the trail and disappears.

A wood chuck is later surprised away from its den and races to the mound of earth before the entrance. There it pauses and switches its short, hairy tail back and forth before whistling sharply and diving out of sight.

A gray squirrel lopes leisurely along a narrow trail through a hedgerow, unaware of a man's motionless presence. Not until the squirrel is almost to a man's feet does it realize that something is different. It hesitates a moment then moves off to one side and continues on its way.

This is only one of a great many close encounters of this sort that a man has had with wildlife of many species over the years. If the wind or some errant breeze does not blow towards the animal and one remains absolutely motionless and does not gaze directly at the animal, then the animal will usually not detect the alien presence or the scent and will continue on its way unperturbed.

Although challenging at times, wildlife observing is always rewarding inasmuch as the observer becomes a small part of the creature's life for a short span of time and thus is able to learn and understand its habits and way of life. A live animal is much more interesting than one that is dead.

As summer deepens, a countryman continues to watch and listen, always attentively, as each new actor appears upon the stage of the season.

Clouds hide the face of the early morning sun, yet bright areas in the cloud cover show where it is in the sky. Now and then a few bright rays break through and send shining shafts of light earthward and thunder rumbles ominously within their midst.

The air is yet cool and filled with varied odors, none disagreeable but rather earthy and pleasant. Dead leaves, mature plants and the earth, still wet from the previous night's showers, exude a rather musty smell. At the same time, there is a somewhat spicy, almost aromatic blend of scents, no doubt the combined redolency of the numerous plants and herbs now in bloom as well as the many kinds of matured grasses.

The clouds slowly break apart and the sun appears, to warm and dry out the countryside, which in turn causes the humidity to rise.

A hummingbird, having discovered a patch of milkweed, flies purposefully from flowerhead to flowerhead, hovering before each pink blossom while it partakes of the nectar and the numerous small insects that have been attracted to the blooms.

Several robins pursue a sharp-shinned hawk that lately has been making it a practice to hunt in the area. The robins are very vocal and the hawk darts here and there in its attempt to elude them.

The insistent voice of a pileated woodpecker is heard and the bird is soon seen perched near the top of an electric pole amid the wires and insulators. It appears oblivious to its perilous position and continues to voice its opinion of worldly matters while its crest flashes crimson in the sunlight.

A mourning dove alights on a wire near the woodpecker. Its sudden arrival seems to startle the woodpecker which manages to fly, unscathed, from its dangerous position and off into the nearby woods.

Drab-colored grasshoppers fly up with rustling wings from the dusty vineyard road ahead of a countryman. Shallow cups, formed in the dusty track, confirm that numerous small birds have recently taken dust baths to rid themselves of the ever-present parasites that are the bane of the avian world.

Kingbirds call from their perches atop grape posts in the vineyard. One of them suddenly launches into the air to capture a large insect which it takes with it over into the next section.

Overhead, several goldfinches ride their roller-coasters low over the grapevines, calling cheerily out to one another. Suddenly, they whirl in mid-flight and settle amongst the vines to continue their conversation.

The day heats up and all afternoon the indigo buntings sing from the edges and open areas. A few summer insects tentatively try their instruments in preparation for their upcoming concerts.

Shadows lengthen as the day star declines westward. Already the yellow flowers of the evening primrose have opened. As dusk veils the landscape a hawk moth hovers before one of the golden blooms and uncoils its feeding tube to sip nourishing nectar.

From the now darkened woodland a wood thrush calls "pep-pep-pep". A few moments pass and it calls again, a disembodied voice calling from the darkness. Then the sweet carillon call of "ee-o-lay" is heard several times before the thrush settles down for the night, bringing to a close another mid-summer day.

Many things come to mind as a countryman contemplates the month of August. There are the hot, hazy days and the warm humid nights with not a breath of air stirring, or so it seems. The occasional thundershower relieves the dry earth but does little to alleviate the oppressive heat and after it is over it only adds to the humidity. But then, this is what we waited all winter for. We need to learn to live with not strive against this wondrous world we live in.

Goldenrod blossoms are appearing on the stage of the seasons, summer's golden scepter, the royal baton that she will wield beneficently for many weeks to come.

Fields of grain come to fruition under the summer sun; some are already ripe for the harvest. Hayfields wait patiently for their second cutting to be made, and green fields of corn, tall and tasseled, wave their leafy arms to each passing breeze.

Thistles raise their prickly stalks along the roadsides and in the fallow fields, while their dark green forms dot the pasturelands where the cattle have grazed around them. Their lavender and purple blossoms will soon mature and develop seeds, each possessing its own tuft of thistle-down which may carry it far from its parent plant.

With the blooming of the thistles and the goldenrod comes the slow but steady decline of the summer vacation of the wild canary or goldfinch. Up until now they have been enjoying their joyous and carefree life. Now they must soon turn their attention to nest-building and the rearing of the next generation.

Katydids appear here and there. They are still small and growing each day towards maturity. They are as yet miniatures of what they will become in another month.

Blackberries are a sure sign of the maturity of summer. The large plump fruits ripen quickly in the hot sun. A man thinks this is one of the best harvests of the year, and no matter how hot or how humid the weather he is sure to be afield to gather his share. He will be content to have some with his breakfast and enough for at least one large, deep-dish pie.

One particular patch of berries has its own guardians. They are paper wasps of the open-comb variety. A man moves carefully about the nest as he gathers the dark fruit, cautiously picking near the nest and always attentive to their behavior. He is respectful of their habitation and they are deferential in return. A man is of the opinion that it is a fine thing to be on good terms with Nature's children as much as is possible.

More and more of the birds are taking on a rather ragged appearance as they commence their summer moult. Crow feathers are very evident upon the ground these days. Those seen flying overhead usually have one or more feathers missing from wing or tail.

While the other songbirds are losing their feathers, the goldfinches will keep their fine suits until their young are reared and then, almost overnight, or so it appears they have exchanged their black and gold finery for a suit similar to that of the females. But for now, they find gold with black trim quite becoming.

The lazy, hazy days of summer often start with early morning fog cloaking the valleys with a mysterious gray shroud.

Flying at tree-top level, a small flock of geese talk to one another as they move from their overnight resting place to some area where they can breakfast in a congenial atmosphere. Their wings beat in perfect synchronization, carrying them soon out of sight and sound.

The farm pond is silent now; only on rare occasions is the silence broken by a frog voice. Dragonflies dart over the surface intent on capturing their prey and barn swallows dip down now and then also.

A doe feeds at the edge of an alfalfa cutting and lifts her head as a man approaches. When he stops, they each eye the other momentarily before she retreats to the shelter of a nearby woodlot, her white tail waving a farewell. As she reaches the woods, a flicker greets her heartily.

An immature hooded warbler flutters after insect prey among the branches of a maple and a short time later a warbling vireo is seen examining the leaves and branchlets of a willow for food.

An airborne bit of orange flame crosses an open space and a cheery voice announces the presence of an oriole. Its continuing calls are reminiscent of late spring and early summer. Their brood raised, the orioles can now enjoy a vacation until it is time to return to the tropics.

The day warms and the humidity increases, but a gentle breeze blows through the branches of the red pines where a family group of titmice call back and forth in nasal chickadee-like tones as they too pursue a meal.

Farther afield a countryman discovers some catnip whose blossoms are starting to mature. He crushes a few leaves between his fingers and savors the redolency that fills the nearby air. Catnip and yarrow are two plants with aromatic leaves that a countryman truly enjoys. To him they embody what summer is all about, yet summer is this and so much more.

Summer is jewel weed growing rank in shady, damp areas its yellow-orange-spurred blossoms speckled with darker color drooping from slender branchlets. Commonly known as touch-me-not, the flowers mature into small seed pods that explode when touched, scattering seeds in all directions. Some later take root, perpetuating the species, while others are eaten by several species of birds. The crushed leaves and stalks are reputed to be a treatment for poison ivy, while the nectar of the flowers is much favored by the hummingbirds.

The milkweed plants exhibit varying stages of development. Those in shaded areas still have conspicuous flowerheads, those which have more sunlight have developed tiny, pear-shaped pods while those out in the open have much larger pods but are still far from reaching maturity.

As he walks among the many stalks of milkweed, a man searches diligently above and below the leaves for monarch larvae, but none are found. A monarch flutters about the blossoms and later is joined by a yellow swallowtail butterfly. A few milkweed beetles scurry out of sight and a ladybug crawls slowly beneath a blossom . . . but no caterpillars crunch their way through the leaves.

Walking slowly away, a man contemplates the future. Will there be any monarchs next summer in time to lay their eggs for a future generation? What about the other species that are disappearing, such as our frogs and toads and numerous once-common species of birds and butterflies? We must do all we can to ensure that our grandchildren have something to enjoy.

An evening ramble in late August may be as varied as the days are. There are subtle differences and then there may be a wide latitude of change. No matter the variance, each evening is unique in its own way.

When there are only a few cumulus clouds and thin, wispy layered clouds in between, the setting sun paints the sky with pastel shades of pink, orange and yellow. The last rays of sunlight causes the outermost edges of the cumulus clouds to glow brightly as shadow-fingers stretch out over the land.

As darkness thickens, the sky remains relatively bright, reluctant to let go the remnants of light from the now invisible sun. Slowly the light diminishes and the evening star appears in the southeast while a bright half-moon hangs in the western sky. As the sun before it, the moon follows its course and as the earth rotates the distant trees hide its beaming face.

In the shadows of the willows, a deer feeds along a vineyard headland unaware of a man's presence. The evening breezes flow from the deer towards the man and he remains motionless, watching. The deer continues to crop the grass, occasionally lifting its head to look for danger. Her tail continuously flicks at biting insects as she slowly moves down the headland, still unaware of the man.

Some evenings the moon is veiled with thin clouds that thicken as time progresses. There is no starshine this night and soon even the moon is hidden from sight. Streamers of fog move across the countryside and the distant ridges already are obscured.

Sheet lightning briefly creates luminescent areas here and there in the darkened sky adding to the mystic qualities of the fog.

Fireflies twinkle in the moisture-laden air adding yet another dimension to the night.

And then there are the insects. The air pulsates with the vibrations of thousands of tiny wings rubbing together. Some have rasps and files on their legs and together these form the instruments of the vast orchestra that plays to the summer nights.

Along roadsides, in the fields and about the dooryards, crickets can be heard playing their diminutive fiddles in a similar manner. A man likes to hear the chirps of crickets. There is a friendly sort of quality in each squeaky note, and on a somber evening like this they lend a sense of cheerfulness to the atmosphere.

A small bat swoops low over the vineyard, rises quickly and disappears in the darkness. Higher overhead, a larger bat flutters across the open space between two stands of trees, momentarily disappears then reappears on the other side, quickly rises on leathern wings only to swerve abruptly after a dodging insect, swoops once more to a lower level before vanishing in the darkness.

The mosquitoes have discovered the man and found him and the moist evening air both very agreeable, especially the former.

Eager to leave the pestiferous insects behind, a man walks quickly homeward. He stops a few moments at the old rusted barrel to look in at the wasps that have built their open house just beneath the rim. Looking closely in the darkness, he can see them huddled together on the grayness of their paper comb, each sharing the other's warmth against the increasing coolness of the night. And the little lanterns of the fireflies flash dully in the thickening fog.

September – School – End of summer: it all seems so final, yet in reality it is the beginning of many things. It is the beginning of cooler weather, a commencement of learning and the maturity of many things.

Signs of this maturity are everywhere to be seen, in the woods, the fields, the orchards, the vineyards and most assuredly in our gardens.

Bluejays have discovered the maturing ears of corn and they are not very quiet in their pilfering. But then, when were jays known to be quiet except when raising their young? Now that all of that is over for the year, it is time to be the boisterous, quarrelsome, vociferous birds that they are known to be.

One of the local blue herons flies slowly overhead, its long neck tucked neatly back against its shoulders. No doubt it is visiting among the many farm ponds that are in the area as well as the numerous marshes.

While taking an early morning walk, a countryman hears the voices of house finches as they investigate the ripening cones in a tall arborvitae tree. Apparently they are finding them to their liking as they busy themselves with the cones as the sunlight sends its warming rays over the land.

The swallows are seen less and less these days. They are enjoying a welcome vacation from familial duties. Soon they will be gone for another year and when their cheery notes are again heard the world will be just nicely turning green for Spring.

A white-breasted nuthatch and his family are exploring the limbs and trunk of a maple, their sharp, nasal calls to one another assuring a man that there will be nuthatches around for him to enjoy.

Without the continuous help of woodpeckers, nuthatches, titmice and chickadees, our forests and woodlots and door-yard trees would not survive the many decades they now do. These birds are busy year around, ever seeking out the eggs, larvae and the resultant adult insects that, if left alone, would soon destroy our trees and shrubs.

These birds depend on standing dead trees and dead limbs as places to excavate their cavities for shelter and to use for nesting.

Each year these bird species, as well as others, find it more difficult to find suitable homes due to the constant changes in their environment because of the increasing activities of man. Building and putting up bird houses for specific species will help ensure their continued presence and the survival of our trees.

The wheel of the seasons slowly turns bringing with each new day some new insight of the world of nature. A man welcomes September just as he will welcome each month in its turn, for each has much to offer. He will welcome it as a month of ripening, a time of fulfillment and maturation. This is the season when one can begin to see the end results of his spring-time labors and the final fruition of all Nature's endeavors.

The hot, sultry summer days are drawing to a close, and already there is a touch of Autumn in the air and the surrounding countryside. The early morning dew lies heavy on the grass and shimmering silken sheets are spread here and there on the grassy carpet. Funnel Weaver spiders have spread their thin webs and wait below the small funnel woven near the center and wait patiently below for some insect to become entrapped.

The morning fog slowly lifts and dissipates, revealing treasures in the fields and along fencerows and beside the roadways.

Treasures of gold stretch for acres in many places, as the fresh new blooms of goldenrod blanket the fallow fields, hiding the ripened grasses and reflecting the golden rays of the summer sun.

White New York asters, that have decorated the countryside all summer, are now being supplanted by pale amethyst gems of lilac-tinted asters.

Here and there other floral gems brighten the landscape as the deep purple of New England asters unfold, one by one, from their numerous buds. These royal amethysts have, at their centers, a round nugget of old gold, and as each bud unfolds, the stem is daily increased with unsurpassed beauty.

Goldenrod and asters, Queen Anne's-lace forming her mature bird's-nest shape, all proclaim the closing of a season. The yellow and orange blossoms of butter-and-eggs reach above the grasses in the mowings about the vineyards and some of them find their way to a man's home to decorate a table.

These are some of Nature's more colorful treasures she so willingly shares. Others may not be so conspicuous, but they are just as important and just as interesting.

For example, the spruce cones become larger every day, slowly lengthening and expanding as they grow towards maturity. Now, beginning at their tips, a ripening brown displaces the tender green as it creeps upward. Globules of resinous sap decorate many of them and they glisten in the sunlight.

Gray squirrels scramble and scold in the walnut and hickory trees, each declaring that the other has over-reached his territorial boundaries and demanding immediate retreat.

This is the time of year that a man takes a container and searches out the dark purple clusters of elderberries. The birds have already taken their toll and there is scarcely enough left for a small pie, but they are sufficient.

A tree frog chirps, and by the sound a man thinks that it is one of the spring peepers working its way to a marshy, muddy place to spend the winter. A man picks berries and listens to the little ventriloquist.

Walking homeward along the dusty roadway, a man steps carefully so as to avoid stepping on the numerous juvenile crickets that hop and scurry out of his way. Out in the grasses, adult crickets fiddle away the warm September afternoon.

A countryman feels as rich as any king might, as he enjoys all these treasures of sight and sound, and the treasures of the elderberries in his hand.

He once more counts his blessings and gives thanks for the bounties supplied by the beneficent hand of the Creator.

The sights and sounds of approaching Autumn are everywhere in evidence. Some of the local geese have banded together into a small flock of about twenty-five or so and made the early morning air resonant with their clamorous calls. This indeed reminded a countryman of the fast approaching season. All too soon their northern cousins will be making their way southward.

The sumacs are unfurling their banners of orange and crimson while beneath the big-tooth aspen tree large yellow leaves lie on the ground like golden pages from summer's book.

Numerous monarch butterflies now flutter about the countryside, their fresh, rich colors indicating that they have just recently emerged from the mint of the milkweeds. A countryman is pleased to know that there is, indeed, a new generation of monarchs to take the places of those which have fallen by the wayside.

Deep yellow sulphur butterflies are equally conspicuous as they flit along the roadsides and across the fields, stopping here and there to rest and daintily sip nectar from the many late summer blossoms.

A man stops at the old barrel to see how the wasps are coming along. He was concerned how they had fared during some of the hard rains that had recently fallen. All appears to be well and he notices that the paper comb has increased in size since the last time he had looked in upon it.

Two, sometimes three, wasps fan their wings almost continuously from their stations at the top edge of the comb, providing a circulation of air across the nest. This seems to be a sufficient mode of air conditioning.

Several newly emerged wasps cling to the fragile paper structure, taking their place alongside the many others that now cover much of its surface. Numerous closed cells indicate that there will be many more wasps emerging before the season draws to a close.

A man watches the activity about the nest for some time. Meanwhile, several white-face hornets come and go with material gathered from an old fertilizer sack draped over the edge of the barrel. Even Nature's children recycle what we humans discard.

Out of the corner of his eye a man sees movement and hears a slight sound. A deer mouse scurries up a piece of old exhaust pipe protruding above the barrel rim, its slight form quickly disappearing into the upper end.

A man continues to observe the comings and goings of the hornets and the ceaseless activity of the wasps. Just before he turns to leave, a very slight rustling sound attracts his attention. The little gray deer mouse has scurried back down the exhaust pipe and disappeared into the depths of the trash in the old barrel.

It is truly amazing, he thinks, just how much can go on in such a small area and in such a short span of time. There is always something happening in the natural world but we humans are not always aware of these things. Then too, we often are unable to be at the right place at the right time. But if we learn to use all of our senses, we may well be let in on a goodly number of Nature's secrets.

# AUTUMN

Autumn, that wonderful and gloriously beautiful time of year has arrived, bringing with it a touch of sadness as little by little one sees the many birds of summer depart for their winter homes. To a countryman the sadness comes with the passing of the wild geese on their southward journey. When they have gone one had better have things pretty well ready for winter.

Just now, as the days and nights have become equal, there are still many mild and mellow days ahead, days filled with bright sunshine, bright autumn flowers and, finally, brightly colored leaves sifting softly earthward.

The clouds, the wind and the very atmosphere all affirm that summer's coinage has been spent and that there are only a few golden coins of Autumn left and that they should be disbursed wisely.

The local geese have become increasingly restless and can be seen and heard most every day about and between the lakes. Although they may spend the whole year in the area, yet their inner clocks tell them it is time, time to re-pattern their daily activities and routines. The shortening of the days disclose to all the wild folk that it is time to find winter homes, find and store away food for the long, lean months that lie just around the corner.

The bluejays proclaim that autumn is here as they become more vocal while they commit petty thievery in the cornfields and squabble among themselves. The jays now vie with the squirrels for ripening acorns, relocating many an acorn that had been wedged into a limb-crotch by an industrious squirrel.

With cooler weather, the wasps in the barrel seldom need their air-conditioning system. Soon their activities will draw to a close for another year. Their nest has grown in size and their numbers have increased but only a very few will find a hideaway and survive the coming cold.

A countryman is always pleased to unexpectedly meet old acquaintances among the wild folk. Such was the chance encounters he had one day. Quietly moving about the underbrush was a rufous-sided towhee or chewink. There was no loud scratching of leaves or hearty call this day, just a few brief glimpses through the leafy screen.

Then a red-eyed vireo appeared momentarily in a nearby willow thicket and was soon joined by a pair of chattering wrens. A northern yellow-throat watched from behind his black mask and uttered a few sharp call notes.

All too soon most of these friends of field and forest will have gone to warmer climes. Then the woodland birds that have been discreetly raising their families all summer will start to visit around the dooryards and at the feeders. Birds that have spent the summer in the sub-arctic will also be arriving to cheer us with their varied voices.

Autumn – a countryman thinks that even the word has a ripe, mature sound to it. That is what the season is all about. The maturity, the ripeness and a whole world full of bounty waiting to be gathered in and stored away.

By early October the vineyards fill the air with the redolency of their ripening fruit. Soon they will be the scene of much activity as the harvesters move up and down the rows and trucks arrive and leave again with loaded bins.

Small flocks of geese already are moving in a southerly direction, their honkings indicative of things to come. It would make one wonder if they sense a cold weather system coming or perhaps they just want to get an early start before the main migration commences.

The spruce trees are loaded with cones this year and several gold-finches busily separate the seeds from their protective coverings. The discarded pieces float slowly earthward, glistening in the sunlight.

The finches converse musically among themselves as they continue to feed. The black and gold of the males has faded and soon will disappear completely except for their wings which will remain a little darker than those of their demure mates.

Cardinals are out and about now that their moult has been completed. They are back at the feeders and more conspicuous afield as they feed on the many varieties of wild berries that are now ripe.

The mew of a catbird comes from a thicket where it also is enjoying the ripe fruit, and the churring call of a red-bellied woodpecker rings out from the big willow tree where it has been poking and prying for insects and larvae in and beneath the deeply furrowed bark.

Prickly stalks of teasel stand brown and sere among the showy blossoms of goldenrod and asters. Their seed heads are full of loose seeds which spill copiously to the ground when the heads are jostled or inverted. It seems like only yesterday that these same heads were blossoming with their lavender belts of florets.

Hawthorne trees are adorned with their bright red or red-orange berries and patiently await the various birds that will visit and partake of them. Even after they become shriveled by the frosts they will provide sustenance for birds and mice during the cold winter months.

Golden garden spiders or Argiopes stretch their handsome and intricately woven webs between weed or flower stalks and hang upside down on a little pad of silk zig-zagged near the center. Golden yellow with black markings and long, dark legs, these are the more conspicuous of the orb weaver spiders.

Earlier, while picking blackberries, a countryman would, on occasion, discover a small one hanging on its tiny web. Now they are large and colorful, conspicuous on an equally impressive web. The females will soon conceal thousands of eggs in a silken sac and then die. The next generation of orb weavers will emerge the following year and spend their summer in the sun just as previous generations have done over the millennia.

A countryman awakened to a rain-washed world. Each blade of grass held its raindrop now scintillating in the morning sun. A slight breeze gently moved the grasses and the droplets glowed and sparkled like so many sapphires, emeralds, rubies and topaz.

The once-brown grass of the mowings were again bright green with renewed vigor as their roots drank in the water, while above, the radiant sun sheds its bright beams from an azure sky swept clean by the broom of the wind.

Water stood in little pools and puddles everywhere while small streams spilled from the open mouths of culverts.

As the day warmed, so did the butterflies. White and yellow cabbage and sulphur butterflies, as well as an occasional monarch, could be seen wrestling with the wind that seemed bent on taking them on a course other than they desired to go. By keeping fairly close to the ground they managed to escape most of the stronger currents of air and were enabled to navigate a course more to their choosing.

Such fragile creatures the butterflies, yet they have the strength and perseverance to accomplish their purposes in life against seemingly insurmountable odds. What an example to those of us of a much higher order.

Peepers, those vocal frogs of Spring, peep their way to their winter quarters in the wetlands and boglands. Their voices, so jubilant and so insistent in Spring are now casual and subdued during this autumn afternoon.

Brightly-hued leaves adorn branches here and there on the trees, forming patches of color among the otherwise green foliage.

Each day Nature adds a little more pigment to the leafy canvas by slowly removing the green chlorophyll and leaving behind the bright carotene colors to form delicate designs throughout the cellular structures.

The great wheel of the seasons has turned to the point where the trees no longer need the chlorophyll to transform the energy of the sun into life and growth. It is time for the trees to rest. It is time for the nourishing sap to retreat to the root systems or at least to a point further down the trunk.

As a man looks across the landscape, over the fields and beyond the ridges, he can see the summation of the seasons clearly delineated. The deep green of the cornfields has changed to a yellow-green and tans often predominate as their leaves lose their color and the ears of corn stand or hang, full and heavy with ripeness. The trees in the orchards stand with drooping boughs laden with colorful and ripened fruit.

Woodchucks eat away the warm afternoon, putting on more fat before retiring permanently for their long sleep. Already, their bodies have token on a roly-poly appearance and they waddle as they run across a field.

The rain-drop jewels, the fragile butterflies, the peepers and the multi-hued leaves, yes even the woodchucks, are all tokens of a loving Creator. From the watery jewels to the lofty trees, all bespeak the wonders of creation.

The sun tried to shine dimly through the rifts in the overcast sky. Darker cloud masses were scattered here and there and the air was cool.

Two freezes and a frost had left their mark on the landscape. The vineyard foliage now hung curled and crumpled, the frosted leaves appeared old and rusted if their color now was any indication. Cornfields, no longer vibrantly green, stood in ranks like khaki-clad soldiers, their tan leaves rustling in the light breeze that blew out of the north-west.

Red-winged blackbirds feasted among the fat ears, putting on fat for their long journey south competing with the jays for the ripening kernels.

Wooly Bear caterpillars wandered here and there headed for places best known to themselves. There was twice as much black on the front of their bristly bodies as on the rear. This is supposed to be an indication of winter being harder the first part of the season. Conifer cones hang in clusters and walnuts are in abundance. Time alone will determine the seasons' length and severity.

The wasps huddled together for warmth on their nest. When a man cautiously touched them with a forefinger they could do little more than raise their posteriors defensively and feebly buzz their agitation at being disturbed. As the clouds slowly moved to let the sun warm the nest, they too would warm and become more active and a man would not dare to take such liberties.

Flocks of starlings swirled and wheeled overhead and finally alighted on an open area and commenced foraging in orderly ranks.

The bright fields of goldenrod have faded with maturity and the sharp fingers of frost. The fields now are tinted with tans and brown shades and deer can stand inconspicuously among the weeds and grasses.

Song and field sparrows and goldfinches chattered and twittered happily among themselves as they feasted on the abundant supply of ripe seeds.

A dragonfly coursed over the field and a bumblebee buzzed about a clump of goldenrod still brightly blooming from its place in a sheltered corner where Jack Frost had not stopped. Further back in the corner a catbird mewed softly to let a man know that it had not yet departed.

The soft, melodious calls of bluebirds mingled with the tenuous caroling of a robin perched in a big hickory. Somewhere, hidden in the still-green foliage of the hickory, a gray squirrel scolded and sputtered, but the robin paid him no heed and continued softly calling.

Two nuthatches chuckled nasally to one another as they climbed about the branches of a nearby walnut, then flew to the big willows to continue their search for insects and larvae. A woodpecker briefly called nearby.

High overhead a red-tailed hawk circled, the sun accentuating the color of its chestnut-colored tail as it dipped and turned on the thermal. Lower down, a turkey vulture soared and tipped effortlessly as it searched the ground for a well-ripened meal.

The months of March and October have a lot in common. Both have warm, mellow days when robins and bluebirds, red-wings and song sparrows voice their opinions of the seasons. These are both times when the summer birds are absent and the winter birds are more conspicuous.

A man thinks that October is to the end of summer what March is to the beginning of springtime. Both hold promise of things to come. Both bespeak the wonders of their Creator and all His vast and marvelous creation.

Color flows over the hills, along the ridges and spreads down to the valleys and across the lowlands.

The oaks, white, red and black, provide an array of hues, mainly tans and browns, yellows and magenta and deep reds. The hickories and walnuts add their shades of yellow and yellow-green. Crimson and scarlet adorn the swamp or soft maples and their foliage flames in the moist lowlands. Along the village streets, down quiet country lanes and all across the uplands the sugar maples flaunt their manifold shades of red, orange and yellow.

Contrasting in a very pleasing way are the deep greens of the hemlocks and the pines as they blend their verdure with the many vibrant colors that surround them.

Borne before the autumn wind, the leaves sail, drift or spiral earthward to settle comfortably into quiet resting places. In the woods they are steadily building up a fresh new carpet of color on the forest floor.

Chipmunks scurry across the brightly-patterned carpet, their cheek pouches bulging with seeds, acorns and nuts as they busily fill their storehouses for winter. The woods echo with their chucking calls as they pause to announce territorial boundaries from atop stumps, rockpiles and stone walls.

Squirrels scold and chatter as they too vie with one another for the choicest food supplies. They dig shallow holes in the earth beneath the leaf litter, carefully place a nut or acorn in the hole and then carefully tamp the dirt back over it with patting movements of their front feet.

The loud, sharp call of the pileated woodpecker is heard long before it appears overhead, flying with steadily-beating wings and undulating flight. The sunlight glints on its crimson crest as it nears the big willow trees. It veers downward and alights on a small limb. How far it has flown is unknown but it appears tired. It rests on the branch for a few moments before continuing on its way, its call following it into the distance.

Song sparrows glean the tiny seeds from the gray fluff that now adorns the stalks of goldenrod at the edge of the meadow. White-throated sparrows peck at the dirt beneath vine-covered bushes in search of bugs and worms and whatever seeds may have fallen down from above.

As a countryman walks along a roadway, several juncos fly up, their outer tail feathers flashing white as they disappear into the still-green undergrowth.

The year is becoming old and somewhat the worse for wear, but it seems determined to have one final burst of glory before it moves forward through the cold glittering gate of Winter.

Several days of wind and rain wash and scour the countryside. Streamlets and brooks are alive once again, sometimes gurgling, sometimes gushing and always flowing towards some distant region.

The dry, dusty earth is now mud in many places, creating a multi-paged record book to be signed by such passers-by as deer and raccoons. Perhaps a fox or even a wild turkey will leave their autograph.

The forces of nature have disrobed many trees such as hickory, walnut and ash and cherry. The tops of many others are now naked to the elements while full skirts of leaves cover their lower limbs.

A thickening carpet of leaves covers the woodland floor and lies beneath each outlying tree as well. Their wet forms exude a rich, earthy and often spicy odor that fills the autumn air with its fragrance.

While the peak of flaming foliage may be over, yet the landscape is colored with pastel shades of pink, orange and yellow and subtle shades of green and yellow-green. Each phase of the season has a loveliness all of its own.

Singly and in glorious showers, the leaves relax their hold on their branches and fall earthward, scattering themselves like a colorful blanket over the earth's bosom.

In the distance, a squirrel scolds about something real or imagined, while closer by is heard the rasping sound of squirrel teeth opening a nut to disclose the sweet meat hidden within. They work so hard for so little it seems, but one never hears them complain.

Crows dislodge a red-tail from its perch in a tall old maple and it sails out over an open field where it soon out-distances its pursuers. It circles about several times before proceeding to some distant area where its hunting activities will not be disturbed.

The voices of the birds of summer are stilled. A few robins remain and their sharp chirps are heard now and then. A yellow-rumped warbler is seen as it quietly moves through the area, bound for the tropics. This species is the first to arrive in late spring and the last to leave in the autumn. Their arrival and departure heralds the changing of the seasons.

Several chickadees converse with a nuthatch who in turn chortles at some amusing turn of the conversation, and then breaks into hearty laughter as it flies to another tree to continue its search for food.

The closing days of October may be bright and cheery or they may be a somber reminder of days to come. At any rate, a countryman is bound to make the most of them, come what may, for well he knows that all too soon the icy blasts of winter will sweep out of the northland and across the fields and ridges bringing to the earth its well-deserved rest for another year.

It was one of those mild, late October days that a countryman chose to wander a bit farther afield than usual. The sun warmed the autumn woods and a light breeze tugged at the unfallen leaves and scampered through the cornfield, rattling the dry leaves as it jostled its way through the tall stalks. The breeze was in a playful mood this day and not at all rambunctious or rowdy.

From its seat on a rock slab, a chipmunk slowly chucked, its voice echoed by another hidden in some undisclosed spot nearby. The little striped ground squirrels seem like woodland sprites or elves as they scurry and scamper through the leaves and pause now and then to voice their opinion of the days' events. Sun-filled, warm autumn days and chipmunks always seem to go together. They are indeed an integral part of the season.

A man pauses beneath an old apple tree to sample one of the gnarly fruits lying beneath it. Considering the many years of neglect, the flavor of the apple is satisfactory. A man has always enjoyed sampling fruit from the old orchards and wayside trees. Perhaps it is because his father took him rambling at an early age and introduced him to the various fruits of the seasons. Well can a man remember those early jaunts afield and the many adventures father and son had together.

Several crows met in the corner of a field to discuss some topic of mutual concern, while a jay flashed blue and white through their midst, disrupting their conversation momentarily with its raucous cries and presence.

Following the edge of a cornfield, a man finally arrived at a small woodlot and entered, using a deer trail for passage. Deer often provide suitable trails for one wandering the woods and brushlots.

Halfway through the woods, a movement is seen farther ahead. Pausing, a man glimpses the dark body of a wild turkey sneaking warily through the undergrowth. The beard adorning its chest proclaims that it is a tom.

Along the north edge of the woods in the clearing beneath the power line, several stalks of great mullein rise majestically from masses of huge velvety, basal leaves. One of these towered seven and a half feet upwards.

Off to one side, four stalks of goldenrod gloriously bloomed, each one attended by its own bumblebee busily gathering some of the last nectar of the season.

Several small clumps of golden cinquefoil bravely bloomed along the trail a man used on his way home. This could well be one of the last of the really warm days of the season and it was a day that a countryman would savor to the fullest.

November robins are a joy to hear. Often their voices proclaim the essence of Spring rather than the arrival of Winter. Early in the morning they often chirrup and sometimes carol, taking a man back to late March and early April when their voices dispelled the weariness of winter. Sometimes the bluebirds join them in warbling as sweetly as when the colts'-foot bloomed yellow along the roadside ditches.

The spell is broken by the small flocks of geese winging their way southward, their numbers decreasing as the month wears on.

Reluctant to leave its summer territory, a blue heron still roams over the countryside from pond to pond. One day, a great egret invades the area, its ghostly white form etched against the autumn landscape as it flew in the typical heron manner, neck tucked back and long dark legs trailing out behind.

From their perches in an oak, cedar waxwings take turns plucking frosted grapes from a nearby wild grapevine. A robin joins them for a few minutes and samples a few grapes before continuing on its way.

A yellow-rumped warbler flutters from branch to branch on its way south while a troupe of chickadees do acrobatics nearby, conversing among themselves. In an adjacent tree, a pair of downy woodpeckers forage together, yet separately, each taking a different area of the tree to closely inspect before flying off together to do the same thing all over again.

Clouds move in from the west and scattered flakes of snow sift listlessly downwards. This is the first snow of the season and man knows that it will not last, but a winter-like chill pervades the air.

Before he notices them, a man hears two deer rush for cover. Instantly, he stops. They had been in the field next to a woodlot before they had betrayed their presence. The doe peers nervously from her semi-concealment in the brush, while the buck stands just inside the edge of the woods. The wind is blowing from them and the buck's quivering nostrils cannot pick up any alarming scent, yet he knows that a creature of some sort is nearby. Alert, his eyes and ears work overtime trying to determine what disturbed them. The air is charged with his intensity. Several minutes pass. Then the buck relaxes a bit and nibbles tentatively on a nearby leaf, then casually rubs an antler tine along a branch. The tine and seven more are attached to the main beams, forming a splendid rack of sharp points, polished to the color of old ivory. He takes a couple of steps forward and looks about.

How regal he is and how powerful, his neck swollen by the hormones surging through his system at this time of the year. How easily, a man thinks, he could put down an unarmed human if his latent fear of man should suddenly be dispelled. Those bulging muscles and sharpened tines could be formidable weapons indeed.

Man is the intruder and this is his home, perhaps even his bedroom, and this is one of his mates to protect. How often mankind thinks in dominant terms concerning the creatures around him. How often he forgets that they have their lives that are also filled with emotions and desires often similar to his own.

The doe moves off at an angle and the buck once more turns his attention to her and slowly follows a little behind and off to one side until they disappear from sight and their soft sounds are no longer heard.

The snowflakes float softly down and settle on the soft seed heads of the goldenrod and asters. It is not winter yet, but it is coming.

Fall rains sweep over the land, filling wells, farm ponds and creeks, and an occasional cellar. Rocky-bottomed creeks that have been exposed for several months now sing happily as abundant water flows over the stones. The fluid melody will be heard until it is finally hushed by the icy coverlet that will eventually spread across their singing surfaces.

Snow squalls replace the rain as the temperature drops and the wind shifts from the south and brings icy blasts from the northwest.

The wind and rain have done a rather thorough work of taking the leaves from the trees and now their shrunken and withered forms will provide a warm blanket for root systems and tender plants.

Clusters of leaves still cling to some of the oaks and the beeches. Leaves on the smooth-boled beeches will cling all winter until the new leaves push them off next spring. Not all the leaves will stay on all winter, but a goodly number will, their crisp, dry forms rustling in the winter winds. The leaves whisper softly until the wind intensifies, then their voices are heard in brief bursts of exclamation and in persistent periods of very animated conversations. They are able to catch their breath when the wind subsides, but are ready for more dialogue as soon as the wind returns.

A countryman finds enjoyment in listening to the wind as it conducts the woodland orchestra of leaves, branchlets and boughs. There is a hint of sadness to this symphony yet there are joyous chords that blend into a tranquil harmony that one finds pleasing to the ear and quieting to the soul. Sympathetic chords are struck that prove soothing to taut nerves.

The recently arrived white-throated sparrows are now being joined by their white-crowned cousins and, together with the juncos, they make melody in the thickets, their varied voices bringing notes of cheerfulness to the gray mid-November day.

A man watches and listens. As he watches the juncos, he notices that one is colored a bit differently than the rest. Upon closer examination, he discovers that it is one of a western race that he became intimately acquainted with during his sojourn in Oregon. A brushing of brown along the back and a buffy wash along each side distinguish this visitor from the distant West. Only on rare occasions do they find themselves this far east. A man watches the once-familiar form as it scratches with its fellows.

A wild turkey is disturbed in its feeding along the edge of a corn field by the approach of a countryman. The turkey runs rapidly for several feet, then flapping its wings vigorously it takes to the air. Soon it sets its wings into a downward curve and sails out of sight into a dip in the field. Turkeys, pheasants, geese, ducks, squirrels and many other forms of wildlife find the cornfields abundant with food. What is left after harvesting is never lost. What the larger creatures miss, the mice glean and store away in their granaries.

The sun disappears behind a cloud bank and flakes of snow fill the air for several minutes. The clouds pass and the sun re-appears to momentarily remove the chill from the air before another cloud mass appears to repeat the action once more. April showers and November showers. One of rain and promise, the other of snow and rest for the land. Another year is slowly but steadily winding downwards towards its end.

Mid-Autumn and Thanksgiving arrive about a week apart with Thanksgiving taking precedence. Harvests are over for another year and their bounty stored away safely in their appropriate places. This is a time for retrospection and a time for being grateful for the past seasons.

Already, squall-lines filled with snow and sharp, bone-chilling winds have made their re-appearance upon the stage of the seasons, if only for short periods of time. It is as though they are rehearsing their lines and preparing themselves for the parts they will play during the coming months.

Now that the leaves are gone, homes are revealed that up until now were artfully concealed by the foliage. There are the leafy nests of the gray squirrels that were cool and airy during the hot summer months. Homes of vireos, warblers and chipping sparrows are now outlined against the bare branches of the shrubs and thickets. Then too, there are the carefully constructed homes of the goldfinches and the deep, gray pouches of woven fibers of the orioles swaying in the wind at the tips of slender branches.

Here and there are the wind and rain-tattered remnants of once magnificent mansions painstakingly built by the papermakers, the white-faced hornets and yellow jackets. Once filled with thousands of industrious inhabitants, they now hang empty and forlorn. Often their interiors are exposed to the elements by the sharp beaks of jays and others seeking out remnants of the once huge populations as an autumn snack.

The wasps in the old barrel no longer cling to their many-celled paper comb. Lack of food and sub-freezing temperatures have taken their toll and they lay lifeless in the litter beneath their natal abode. Somewhere, hidden away in a well-insulated spot, one of their queens will survive to start the cycle of life once again in another place in another Spring.

The sweet trill of the summer chipping sparrows has been replaced by the sprightly notes of the tree sparrows. The two species resemble each other in that both have brown, streaked backs and a rusty cap. However, the chippy has a dark line through each eye overlaid by a narrow white strip. The tree sparrow is larger, its russet cap is not so bright, and it sports a dark "stick-pin" on its breast. The tree sparrow is often dubbed the "winter chippy" because of its similarity and its cheery demeanor. They spend the winter gleaning seeds along the hedgerows and across the weedy fields their voices tinkling in the frigid winter air.

The gray November days will soon merge with the increasingly cold and blustery days of December. There will be days when the low-hanging sun will shine feebly and lend a little warmth to the land.

Clouds thicken across the sky as the day draws to an early close, but they are not yet enough to keep the sun from disappearing in a brief but blazing splendor in the western horizon. Night settles quickly over the land and the mercury in the thermometer descends into an area well below the freezing mark.

The morning sun glows weakly from behind a thin veil of clouds, coloring them a warm yellow-orange just above the horizon.

To the west, a pale moon shines faintly through the overcast as the thin clouds thin momentarily then thicken again to obscure the lunar orb.

Clouds, formed in the frigid air by the warmer waters below, rise above the distant lake and drift slowly northward by the southerly breeze.

The world is enshrouded in silver, the result of a recent freezing rain. Pine boughs droop forlornly beneath their heavy burden of ice. The pasture fence is decorated with short, tiny icicles spaced ever so evenly just about one inch apart as though methodically placed there with the aid of a measuring stick.

A man approaches a large spruce and pauses a moment in admiration. The spruce reminds him of an old fashioned Christmas tree, only this tree is decorated with genuine ornaments. Thin icicles, each about four inches long, hang from every branchlet. The snow that followed the rain adds a final touch by coating the branches from top to bottom. Altogether, it is a thing of beauty, decorated by the Creator's hand, a tree truly befitting the holiday season.

A crow, perched atop a tree, sits a silent sentinel over an icy, ermine-covered world. Somewhere the plaintive voice of a bluebird is heard several times then is heard no more. A man wonders how those fragile birds can think to survive in this weather.

The countryside is filled with silence. Gone are the usual bird notes and voices. Even the breeze whispers as though afraid to break the silence. Several juncos fly up from the road-edge into the nearby bushes in silence, their usual sharp twitterings conspicuous by their absence.

The vines in the vineyard shimmer in the sunlight that breaks through a rift in the overcast. Each grape post wears a jaunty tam-o"-shanter of snow atop its thin weathered form as it steadfastly supports its share of the ice-laden vines.

Across a man's path now lies a short stretch of smooth shining ice. He cannot resist that slick surface before him and, taking several quick strides forward, he slides smoothly and easily to the other side. The sensation is exhilarating and he turns and repeats the action.

For several moments he remembers sliding like this as a child and later in life skating as an adult with his children on a frozen bog stream in Maine. But that was many years ago. Yet, the slick surface yields a sensation that is as familiar today as yesterday and a countryman smiles in retrospection. He turns and slides one more time across the icy strip. There are, he muses, certain benefits to be had on this silent, late autumn day.

The great wheel of the seasons has nearly turned full circle. Eleven months ago the land was in the depths of winter yet it had just commenced a fresh new year. All too soon it will be at that point of beginning once more.

The hours of daylight have continually diminished until they are already at their lower limit. The days are so shortened that when the sun disappears each evening it is as though someone had shut off the light and the resulting darkness spreads quickly over the land. There are no long twilight hours, enhanced by the melodic vespers sung by the birds, to be enjoyed now. If, perchance, the temperature has risen above the freezing mark during the day, it drops dramatically into the frost zone with the setting sun.

The whistle of the woodchuck has long been silenced and the cheery chucking of the chipmunks no longer is heard. They have retired to their bed-chambers as have others who sleep the cold months away. Bears may roam about awhile longer until the first real snowfall but they already know where they will spend the winter and when the time arrives they will disappear into their dens where they will be warm and comfortable.

The life that was so much in evidence in the goldenrod and aster, the milkweed and the teasel and all the other plants has withdrawn into the security of the rootstocks to quietly rest until invigorated once more by the gentle warming touch of Mistress Spring.

Insects are now resting in various stages of development, some as eggs and others as larvae or pupae. Others as adults are hidden away in some secluded and sheltered spot away from the lethal effects of sub-freezing temperatures. There they repose until warmth again overspreads the land and they come forth to start the cycle all over again.

Not yet winter and nearly end of autumn, December often exhibits its capricious character. The wind and mellowness of March are here one day while the following day may well be filled with snow squalls and frigid temperatures. Interspersed may be a day of mist and rain followed by an evening of fog that cloaks the land in near obscurity.

This is a time of sudden and dramatic changes, but it also is the beginning of a time of rest. It is, indeed, a time of respite for the land before another season of renewal returns.

Dark, dismal days often occur during the last month of the year. It would almost appear that just because the sun hangs low in the southern sky all day that that is also a good excuse to hide its bright beaming face.

Of course, the sun has nothing to do with it at all as we well know. Nevertheless, as the late autumn days so often are overcast it causes one to think along those lines. At the close of some of those days the sun will peek out through a rift in the clouds just before it disappears for the night, washing the clouds in shades of orange and yellow and red adding a brief touch of color to an otherwise gray canvas.

On a rare occasion the morning will dawn bright and clear. Heavy frost whitens the roof-tops. Frost crystals cling to the weed stalks and crystalline ice coats roadside puddles and the water in the ditches.

On one such morning a countryman cannot resist a short ramble to see if the morning sun has activated any of the wild creatures which have recently been so conspicuous by their absence. His walk is soon rewarded.

Despite the cold and frost, bluebirds are soulfully calling. Some are perched here and there while two more fly overhead and these actually call in a cheery vein, as much as to tell the world that one must look on the bright side of things if one is to be happy. They did lift a man's spirits as he continued on his way.

A half dozen robins sit in the tops of two adjacent trees. Facing the sun, they soak up the sun's rays which warm their bodies, dispelling the night chill that had slowed the life forces within.

Out in a field a male cardinal sits low in a bush. In the sun light, its plumage glows like an ember against the almost black branches amongst which it sits. At the woods-edge a female briefly appears only to disappear into the underbrush.

Two gray squirrels take turns entering and leaving an old crow's nest atop a white pine. The top of the tree is broken off, exposing much of the nest. A countryman muses some moments as to what use the banner-tails are making of the nest. Pondering the situation, he comes to the conclusion that they are utilizing it as a storehouse rather than for winter quarters. It is far too exposed to the elements and predators to make a safe home.

The clamor of geese is heard in the distance. Using his binoculars, a man picks out one flock, a second and then a third. He watches as they draw closer, their formations shifting from lines to vees, wavering back and forth. Suddenly they begin to descend. Many have already set their wings in a downward curve and the rest soon follow suit as the first birds lower their legs in preparation to land in a recently harvested cornfield. Within moments, all the geese have landed. They waggle their tails, adjust their wings to a comfortable position and commence foraging for their breakfast.

As a man watches them he remembers that breakfast of a different sort also awaits him. Crows caw in the distance and jays screech at one another and the world in general as a man retraces his steps homeward. It would indeed appear that the sun had once more worked its magic upon the world and caused the wild creatures to actively go about their varied business.

# WINTER

The days of the year have reached their lowest ebb. In fact they are lengthening a few minutes each week, but at this point it is not only unnoticeable but negligible as well.

A countryman has watched all autumn as the westering sun declined to a position behind a certain tree in the southwest. A man has no need of a calendar to tell him when winter arrives. It is true that the tree is not totally accurate, but the sun never shines beyond that tree. Within a short span of time the sun will start is northward swing and as it does the mercury in the thermometer will sink lower and lower within its glassy tube.

For many decades a countryman has watched, and been a part of, the ebb and flow of the seasons. He has watched and seen how they vary from year to year and yet, in spite of their variation, they pretty much arrive on time.

First frost, first snow, first geese and first robins, all of those often occur in the same week and sometimes on the very same day season after season, year after year. All is programmed to an infinitely accurate celestial clock.

These late December days are the fulfillment of the year's promise. The promises made when the days were at their peak are now realized when it is the nights that are prolonged.

Everything in nature seems to have come to a standstill along with the weak, wan sun. A great stillness has settled over the land as if in anticipation of some great event to transpire. Bird voices are muted to a great extent. The cardinals flash their crimson colors across open spaces and make no comment in passing. Only the crows and the jays call loudly, and then only on occasion.

A sunny day will usually activate the wild folk, for a few hours at least. A long-silent titmouse calls from the depths of the woodlot and a flicker bursts forth with its "clee-up" call. A robin chirrups nearby. Crows, having nothing better to do, harass a hunting red-tailed hawk until it whistles shrilly in sharp protest.

Several deer bound through a brushlot, their white flags signaling their rapid retreat along with the crackling of undergrowth and dry branches.

A gray squirrel disrupts a gossiping group of chickadee as it glides over the branches, a black walnut grasped firmly in its mouth. It stops momentarily as it reaches an open space between the trees. Bobbing its head up and down, it gauges the distance to the nearest distant branch then, bunching its leg muscles, it leaps up and outwards. Alighting on a slender branch, it runs quickly up to where it thickens perceptibly even as the branch is dipping downwards beneath its weight. Gaining the security of larger limbs, the squirrel is once more on its way, the banner-like tail flowing lightly out behind it.

A man enjoys watching the tree squirrels in their travels. He enjoys watching their aerial maneuvers and never ceases to marvel at their dexterity and acrobatic accomplishments. In a sense, they appear to be poetry in motion with a lively meter to match.

Right on the heels of Old Man Winter comes a new year with all the promises and fulfillments of each season that lies ahead. When the leaves fell last autumn, tiny leaf buds had already formed behind them. Now, they are dormant in their protective sheaths yet are replete with the assurance of an abundance of new leaves come spring. Throughout all nature, new life is resting or is slowly forming. Embryonic deer and bear, now in the formative stage, in a few short months, will be born to form a generation of their species. They will run, jump, climb and play, always ready to explore their bright new world. Just now they are forming, cell by cell, and the winter will pass without their knowledge. A countryman reflects, what would the world be like without this regeneration that is repeated year after year? The world would become empty and even man eventually would disappear. There are so many small but important things going on all the time that we take for granted, until they cease to take place or exist.

Mourning doves fly up from a field as a man passes, their wings whistling sharply as they leave. Later, he notices several perched on a power line, their feathers all fluffed out to insulate against the cold. Their voices no longer are heard as they so often were in an earlier season often just before a thunderstorm broke over the distant ridges. Only the whistling of their departing wings is heard on this gray winter day.

Along the edge of a field in the weed patches, tree sparrows talk softly amongst themselves. Their coloration is such that they often are not seen, only heard, until they fly up at a man's passing.

Life and death are uneasy partners in the natural world. One who will spend time in the out-of-doors soon becomes aware of how tenuous the thread of life can be for the wild folk.

Walking slowly along, a man catches sight of movement off to one side. Turning abruptly, he sees that it is a small bird flying off with a smaller bird clutched in its feet. The binoculars prove what was surmised; the bird was a shrike with freshly caught prey.

A shrike belongs to the songbird family but that is where the similarity ends. Although it does not have strong feet equipped with talons, it does have a slightly hooked beak which aids in tearing apart its food. The shrike is mostly light gray with white wing patches and white outer tail feathers. A black mask completes its costume and adds to the sense that here indeed is a feathered bandit or highwayman of the bird world. Its food consists of insects, small birds and mice. These it usually impales on thorns or a sharp twig to hold it in place while it tears it apart. This habit causes it to be called "butcher bird" in some areas where it is more common. Nowhere is it too common so it is never a real threat to the bird population like the sharp-shinned and Cooper's hawks can be at times.

A countryman considers himself fortunate to have seen the little Butcher once more. He realizes that shrikes occupy a useful niche in the order of things and that the world would be a little less complete without them.

There are many days in late fall and early winter when a countryman meanders about the countryside and there seems to be no wildings in evidence. The woods, the brushlots, even the hedgerows seem, to all appearances, to be deserted. For the most part, silence reigns over the land. There are times when the crows even sit as silent sentinels in tree tops or on vineyard posts unwilling to break the silence.

To the less observant, there is nothing to indicate the presence of any life forms, now or in the recent past. This is the time when the utilization of all of one's senses is demanded.

One need not be discouraged from venturing afield even if, for one reason or another, one or more of his senses has become diminished or defunct. The remaining senses are the more sharpened to make compensation for the loss. Therefore, even under adverse circumstances one can enjoy the out-of-doors even if only to a limited extent.

For example, a brief scent of skunk carried by a vagrant breeze concedes the fact that this animal is about somewhere. Droppings beneath a tree are evidence that the tree was used for some time as a perch or roosting spot.

Half-eaten apples beneath an old apple tree give evidence of the activities of deer, squirrels or rabbits. Mice too may play their small part.

Atop rotting tree stumps, nut shells bespeak a squirrel's lunch, while piles of ravaged cones beneath a conifer indicate the activities of a red squirrel.

A fluffy breast feather barred with chestnut brown informs one that a ruffed grouse flew this way and lost it in passing through the brushlot.

Tracks in mud, snow or dust tell of their owners' passage. Feathers molted or caught on sharp objects, hair and fur, even skulls and skeletons, all give evidence of life and death. To the practiced eye, all of these things denote there is or was life in the immediate area.

A countryman stops often to look and listen. That odd-shaped bunch on the limb next to the trunk of the big maple is in reality a gray squirrel sitting immobile with its tail curled over its back to give added warmth to its backsides on this chilly day. Except for its form, the color of the squirrel is nearly the same as the tree bark against which it sits. To the casual observer there is nothing alive there and the squirrel hopes that the hunting hawk also will pass him by.

The faint honking of geese is heard in the distance and betrays their presence although they are hidden from sight by the tall trees. Nearby is heard the nasal "yank, yank" of a white-breasted nuthatch although the little upside-down fellow is staying out of sight on the back side of a tree limb. A thick grove of spruces also hides the chickadees whose cheery calls can be heard as they forage deep amongst the boughs.

This fresh, new year is a good time to resolve to spend more time out afield learning the ways of one's wild neighbors. Their lives are many times lived briefly and are frequently fraught with danger and adventure yet they are lived, one moment at a time, to their fullest extent.

A countryman greets the new year with anticipation and eagerness, anticipating new encounters and an eagerness to learn something new about the world around him.

Although the seasons come and go with regularity yet every one is different, no two are the same. Spring, summer, autumn and winter, over the years each season, distinct as it may well be, will deviate from what one might expect from them.

There are early, warm springs with balmy weather, early birds and blossoms. Then there are the cold, wet springs when one wonders if warm weather ever will arrive. There are hot summers with a fair amount of rain, then there are the dry summers with no rain and even the rivers become shallow streams. There are the early autumns when frosts arrive a week or two ahead of time much to the dismay of the gardener, then again there are those when summer seems to extend into October much to the delight of everyone. Sometimes winter catches everyone off guard by making a brief but distinct appearance around the last of October. It may continue from that point on or retreat until the middle of the next month. Then there are the classic winters that arrive in the latter part of November and continue unabated until early March. These are the winters that our grandparents used to tell about as we sat around the big wood heater in the living room while the wind howled outside and the snow piled into six foot drifts.

No matter the season, each is distinct in its own way. Just now the brooks and streamlets are open with perhaps a thin rim of ice along their edges. A short time ago they were covered over with ice and snow. They will be that way again. Now they gurgle and splash in wild abandon. When the cold again descends upon the land, its icy fingers will spread ice crystals across all the liquid surfaces that are relatively shallow. The ice thickens and as it does the voice of the brooks becomes subdued and under a deep snow cover it even may be silenced. Running under a thin glaze of ice the brook will murmur and tinkle in a delightfully musical way.

A countryman likes to listen to the fluid tones of all the watercourses, large and small. There is something peaceful and serene to their free and fluid sounds. It is relaxing to pause and listen as the water tinkles, splashes and gurgles along its way over stones, around rocks and any other obstacle in the way. A man considers how much a stream or river may sing its traveling song in proportion to the obstacles that lie in its way. Perhaps we could take a lesson from them as we meet the obstacles that beset our paths during the journey of life.

On those rare sun-filled days, the weed patches beside the road are alive with tree sparrows and perhaps a scattering of juncos. Their voices are cheerful in spite of the raw, gusty breezes that ruffle their feathers.

A troupe of acrobatic chickadees makes their way through the trees. A titmouse keeps company with them, its harsher chickadee-like notes heard now and then rather than the more familiar spring calls of "Peter, Peter."

A countryman welcomes a sunny day in early winter just as the wild folk do. He knows it will not last just as he is aware that cold and freezing have a part to play in the order of things.

It is winter, and who can tell how this winter will turn out. One can be assured that there will be a certain amount of wind, cold and snow. This is the natural order of things. Meanwhile, the sun is swinging northward and we all know what that means.

As a countryman walks in the winter woods on a windy day he doesn't expect to see many, if any, furred or feathered folks. Instead, he listens to the many and varied sounds and rhythms created by the wind as it blows around and through trunk, bough and branch and leaf cluster.

Two trees growing close together rub shoulders now and then and as they do so emit a mournful groan as their trunks chafe one against the other.

Various squeaks and shrieks can be heard as the wind moves branches briefly against one another and then they return to their original positions.

Leaf clusters on the branches of a sturdy red oak rustle in light breezes and as the breeze increases they vibrate in rhythmic response to the crescendo and diminuendo of the wind. Rustles increase in tempo until the leaves rattle noisily so as to be heard from some distance.

Leaf-talk varies between species. The thinner beech leaves converse in a lighter vein than the deeper tones of the thicker and heavier oak leaves that cling more closely together near the tips of their branches. The smaller leaves of shrubs and bushes whisper in subdued tones as though relating some woodland secret.

Tree-tops sway vigorously back and forth as the wind moans through their branches and whips them first this way and then that way. At times they may even clash together with considerable force. A man marvels at the resiliency of trees to withstand the many and sometimes violent forces of nature. As the larger limbs collide with one another some of the smaller branches break off. This is one way that nature prunes her trees.

A large limb, rotten at the center, can no longer withstand the forces of wind and gravity and crashes noisily through the lower limbs to land on the ground beneath with a resounding thud.

Pine limbs bend and sway as the wind soughs through their needle-laden branches. Their soulful song is melodic to the ears of a countryman. Indeed, their sighing song is soothing to the soul as is the brief scent of pine resin that wafts through the air.

The hemlocks, spruces and the firs also lend their boughs to the wind creating tonal qualities that are unique to these conifers.

Exhibiting more skill than sense, a gray squirrel decides to travel through the tree-tops on this windy day. It leaps for a distant branch but at the last moment the fickle wind moves it just out of reach. Twisting its body and using its tail as a rudder, the squirrel maneuvers through the air towards another branch lower down. Clutching slender branchlets frantically with its forefeet, it pulls itself upwards toward more stable footing.

Two crows sway rhythmically with the wind from their tree-top perch as they watch the squirrel recover his dignity and composure and continue along his aerial pathway.

The sharp, biting winds of late January are not conducive to much rambling. What creatures that are not dozing in their winter quarters have moved to the depths of what few stands of timber that are left or sought shelter in compact stands of conifers in the many deep ravines that are to be found in the lake country. These ravines can be one of the best places to be on a blustery winter day.

A man has, in years past, often been within the shelter of one of many of these ravines. The bottoms and often much of the sides are cloaked with hemlocks and pines. A man has been quite comfortable in this protected environment while the wind roared through the trees and across the rim rocks above. So it is no wonder that the furred and feather folk find shelter and often food in these places of seclusion.

When the weather moderates as it will from time to time, the wild ones will return to their former haunts but will stay only as long as they can find sufficient food. Food in adequate amounts is extremely important to all during the winter season.

A countryman is content to wander only short distances afield for he too feels the cutting edge of the wind and each winter it seems to be honed just a little sharper. Still, being of an inquisitive nature and having a love of the outdoors, he must venture forth if only to look and listen.

Undaunted by the wind, a chickadee calls in an optimistic tone as he clings upside-down from a maroon seed cone of a nearby sumac. During a brief lull in the wind a downy woodpecker can be heard working among the tree branches, extracting insect pupae and larvae from beneath loose pieces of bark or from its convoluted surface.

Soaring high overhead, a red-tail rides the turbulent air currents effortlessly as it scans the trees and ground for a meal. Woe betide the unfortunate meadow mouse or squirrel that attracts its attention. Cottontails are not entirely safe, especially from a hawk that has not had a decent meal lately. The hawk too must keep its inner furnace stoked in order to survive.

A blue jay cries harshly while several crows stalk about a cornfield in their quest for stray kernels of corn and the possibility of a careless meadow mouse.

After the cry of the jay all is quiet once more except for the wind, the ever-present wind. A man turns his steps homeward toward shelter and warmth. As he turns, he notices small, silvery spots along several branches of the willow brush growing along a ditch. Tiny pussywillow catkins have dared to peek out from their protecting sheaths! Here they are with two more months left before spring. Such audacity, such courage they show, yes, even what optimism at this point in time. While gray clouds scud overhead driven before the winter blast, a man trudges home and he too has gained some optimism from those tiny, silvery catkin-kittens.

When February arrives a countryman almost can sense a subtle change in the season. Not that there is anything tangible yet there is that feeling of change, a tipping of the balances so to speak.

From one point of view, the depths of winter are upon us and we are scarcely half through the season of wintry blast and frigid cold. On the other hand, the days are lengthening more perceptibly, the sun rising earlier in the morning and delaying its descent two minutes or more each evening.

There is a feeling in the atmosphere that Spring is on its way although by the calendar it is still a good six weeks away.

This brings us to another point of contention. If the groundhog emerges from his burrow to take a look at the outside world and fails to see his shadow, we could be in for an early spring. However, over the years a countryman has noticed that it usually makes little or no difference whether or not the woodchuck sees his shadow. Besides, when he arouses from his winter slumber, he has romance on his mind, not weather. There may be many inches of snow on the ground, but if the weather is suitable, he has business to attend to. When his duties of procreator have been accomplished, he will retire to his underground bedchamber and go back to sleep until the weather warms and fresh, new greenery has grown to a state of succulence.

The cold, blustery days of January extend into the second month; there is no doubt about it, winter is far from over. The wind sweeps across the snow-covered fields, the white crystals billowing and dancing before it.

The air currents abruptly change and for a few moments the snow swirls and lifts higher and higher even as the wildly gyrating mass moves steadily forward. Sometimes several of these twirling columns of snow are in motion all at the same time. A countryman thinks of them as “snow-devils” as they are akin to the dust-devils seen in early spring and summer as the air and ground begins to heat. Still, the formation of the one is during the cold time of year while the other is related to the combination of currents and warming ground.

Whatever their cause, ghostly, giddy, whirling columns of snow prance and dance across the fields and about the outbuildings and then suddenly collapse as from exhaustion.

There is much of winter left, but when a countryman sees muddy footprints in the snow leading from one woodchuck burrow to another, he knows then that there are, more or less, six more weeks until Spring.

A flock of over-wintering geese do not a spring make, but the residents are apparently becoming restless for they are making themselves heard and seen more and more as the weeks go by. They fly back and forth between the lakes talking no doubt about the prospects of an early spring or perhaps they are discussing the imminent return of some of their kin from their southland vacation. It will be only a few more weeks before the earlier arrivals will be gathering in large rafts, darkening the bosom of any one of the Finger Lakes. They will feed, rest and converse in noisy groups before continuing north to their nesting grounds.

The geese are not the only ones that encourage a countryman to think about an early spring. The other day he heard the "tut-tut-tut" of a robin which was answered by another nearby. A short time later he heard another one chirping from a clump of sumacs and then one flew to the top of a tall hickory and perched in plain view. This one also "tut-tutted" a few times, accompanied by several typical tail-jerks.

Along the pasture fence several bluebirds perched on the fenceposts scanning the ground for some sort of insect life. What sort would be out and about at this time of year is something that a countryman has yet to discover although he has seen bluebirds feed in this manner during this season many times over the years. That they are successful is evidenced by their continued survival.

In the brushlot, a woodpecker drummed, practicing for his upcoming mating rituals. Two cardinals conversed, the first a man has heard in over a month. A few minutes later a scarlet splash settled in a tall and scraggly shrub, and "purty-purty-purty" came through the crisp morning air. All of this would lead one to believe that Spring was around the corner. In fact, this is the case but just which corner and how far before the right corner is the sixty-four dollar question. There is still plenty of time for abundant snowfall and icy blasts out of the Arctic before Mistress Spring comes tripping across the ridges and down the valleys if she decides to bide her time and arrive more nearly on schedule.

The wooly bear caterpillars had much less black on their backsides than on their front ends last fall and that would indicate an earlier end to winter. But a man has stated before that he observes but takes little real stock in the patterns on the coat of the wooly bear, never-the-less things will bear watching.

The starlings are gathering and voicing their opinions in the early morning hours now. Starling voices are not noted for their musical qualities except when they decide to mimic one of their more talented neighbors; however, as springtime approaches their calls do become a bit more musical than usual and their feather edges are wearing down. Soon they will sporting their spotted, iridescent new spring suits.

A countryman will watch and listen for more signs of spring. He will hope that the winter will indeed beat an early retreat. At the same time he will not be surprised or too disappointed if the next few weeks prove otherwise. After all, the seasons have a way of arriving pretty much on time. And that is as it should be.

Rambling through a small woodlot of mixed hardwoods and softwoods, a countryman came upon a huge old bigtooth aspen. He had seen this tree before during his wanderings and had seen the large opening in one of the limbs which stretched out like a well-muscled arm. The hole was the right size for a raccoon den but it could serve equally well as a home for an opossum or even a gray squirrel. He had never investigated the tree closely before but since there was a small amount of snow on the ground this day, he decided to have a look around the area.

As he approached the old aspen, there in the snow were an abundance of raccoon tracks, some leading away from the tree while others were returning. This was a clear indication of much activity on the part of one or two animals during the past twenty-four hours or so.

At this time of day, and since the weather was quite cold, there was little doubt that the raccoon was slumbering in its snug den. It would continue to snooze away the coldest days and nights and only emerge as the weather turned mild from time to time. The fat that it had put on during the last warm days of autumn would sustain it as it slept during the coldest months. When out and about in a mild spell, it would scavenge whatever it could find, perhaps a road-kill would provide much-needed energy.

By the time Valentine's Day arrives, the raccoon tribe are also in a romantic mood themselves. Since the males are polygamous, they will wander here and there looking for receptive females, much as the woodchuck is wont to do. It is apparent that they do not take family responsibilities seriously at all. The female, on the other hand, is a very devoted and a very defensive mother and will carefully train and care for the youngsters that will be born a couple of months or so later.

Perhaps this old aspen will become home to a lively bunch of bright-eyed, curiosity-filled little ring-tails by the last of April. To be sure, their mother will have her time well occupied caring for such a group of little mischief-makers.

A small stream flows past the base of the aspen. It no doubt serves as a source of water to numerous thirsty creatures, but it is frozen over now. Sometime, during the freezing process, a cottontail and a domestic cat left their footprints to congeal and remain as mute evidence of their passing. A man walks down the frozen surface, past diminutive frozen cataracts and quiet pools. Stopping a moment he can hear the soft liquid gurgle of the water as it flows beneath its icy blanket.

The gentle tap-tap-tapping of a female downy woodpecker is heard in the woodlot. A man watches as she carefully searches for insect larvae among the dead branches and loosened bark. She busies herself for she well knows that the winter darkness will soon force her to seek shelter in her hollowed-out bedchamber for the night.

A chickadee clings upside-down on a nearby pine branch as it too searches for a bit of nourishment for the coming night.

Curled up in a furry ball, nose buried in its bushy dark-ringed tail, the raccoon sleeps soundly, oblivious to the activity in the outside world.

Lake Country gullies have a character all their own. The long ridges between the lakes channel their excess water down these deep defiles to the long, finger-like lakes that lie between them. Snowmelt also contributes to the constant carving process of rain and sediments that cut ever deeper the channels through the soft, sedimentary shale. At times, the water flows in cascades over rock ledges creating beautiful waterfalls that tumble tumultuously down into frothy pools below. These, in turn, may then slide over the smooth shale and around boulders until they reach another point of descent. This process continues until finally the water flows unimpeded into one of the lakes.

Well over a decade has passed since a countryman last entered one of those deep ravines. One needs to be prepared for steep descents and equally steep ascents when negotiating this type of terrain. It requires much exertion and this can be quite demanding, especially for one of advancing years.

A light snow had fallen when a man decided to once more venture along the rim of one of these gullies and eventually descend to the frozen creek below. It was good tracking snow and several species had been out and about, some during the snowfall and some afterwards.

Along the rim, a weasel had wandered here and there as is their way. A shrew had tunneled just beneath the surface, pushing the snow upwards wherever it went, its movements delineated by a serpentine ridge of snow.

Gray squirrel tracks were beneath the oaks as it searched for acorns or anything else edible. Small pockets were freshly dug in the earth in several places evidence that apparently acorns had been hidden there at an earlier time. Mouse tracks laced grass clumps back together from the rim. Deer walked close to the edge, their sharp hooves providing the traction necessary to keep from slipping over the edge.

A steep roadway provided access to the bottom and soon a man was beside the frozen stream. The murmur and gurgle of the water could be heard faintly beneath the ice and snow. An opening in the ice provided deer with access to the water and tracks showed they had been there earlier.

More deer tracks were seen on the other side of the creek and further upstream as well. It was at this point that a man came upon the trail of a fox. But these were not the kind of fox tracks that he was used to seeing. They resembled, in a way, large cat tracks since no nail prints showed. Of course the snow might be of such consistency as to conceal them. The toe and heel pattern was fox, but it was puzzling. It was even more puzzling when the trail led to a steeply slanted log that extended from the creek bed to the top of the bank. The fox had crouched to leap up on to the log, leaving its heel prints in the snow. The log was relatively small and the fox walked casually up and leaped off the upper end, and continued on its way.

After close examination and some measurements were taken, a man came to the conclusion that only a gray fox would leave a trail like this. Gray foxes have semi-retractable toenails and they are known to climb slanting trees wherever it is to their advantage to do so.

Tracking and trailing is like detective work; it is both interesting and puzzling at times and often quite revealing. The best part is that it leads to a better understanding of who your wild neighbors are and how they live successfully in the environment.

With less than three weeks until Spring, there are already definite signs of the change of the seasons. Pussywillow catkins are becoming more conspicuous, the house finches are singing and the cardinals are commencing to sing their territorial songs also. Perhaps they are only practicing but it seems to a countryman that they are ready to perform their full concert any day now.

For the past two weeks robins have been in evidence and have been chirruping most mornings. One even was caroling in a tentative vein and a man thought perhaps it was trying to induce Mistress Spring to arrive early this year. One morning while there was still frost on the roofs, two robins hopped jauntily over the lawn not realizing that the earthworms could not reach the surface because it was still frozen.

One morning when the robins were cheerily calling and the bluebirds were warbling softly and melodiously, the clear voice of a titmouse rang out from the brushlot. Its voice will be heard more frequently as the weeks slip by.

The downy woodpeckers drum on mild sunny mornings, no doubt attempting to find a piece of dead wood with the right timbre and the most resonance so that when they begin to drum in earnest their woodland tattoos will be well worth their effort.

The last of February it seemed that the March Lion had decided to get an early start on things. Perchance it was only a practice session. At any rate, a man is certain that we will hear his roar in the coming weeks as he prowls about the land. Besides, who ever heard of a "February Lion?"

A baker's dozen of crows harassed one of the local red-tailed hawks to the point that it was no longer profitable to soar over the fallow field in search of meadow mice that no longer had snow to conceal their travels. Taking evasive action, the hawk flew erratically towards the woodlot with its black-coated and black-hearted tormentors in close pursuit. Approaching the edge of the wood, the raptor half-folded its wings and dove well within the protective branches of a sugar maple. Several crows perched in nearby trees and casually voiced their opinions of hawks in general. Protected somewhat from the crows by the close branches, the hawk rested, ignoring the taunts and jibes hurled in its direction.

A small flock of northward-flying red-winged blackbirds called "chek-chek" several times. It might seem to the casual observer that they are a mite early, but the fact of the matter is that they are just about right on time.

Still, a countryman can't help wondering if we might just be fortunate enough to have an early Spring. But then, there's that March Lion.

For the past week, the red-winged blackbirds have been sitting in the treetops and calling their cheery "o-ka-lee!" To their notion it is time for Spring to arrive, and for their part, it already has.

Usually one or two in the group will display their bright red epaulets while the remainder is content to call without the accompanying exhibition. After all, Spring is still over a week away, and there is no need to rush things.

A week earlier, when the first vanguard moved northward, the red wings flew high overhead, uttering their "chek-chek-chek" calls. These were the ones going further north. On occasion some would stop and rest and feed a day before continuing on their journey. Then the locals begin to arrive and soon afterwards their gurgling spring voices can be heard.

Recently, a group was in a tree-top conversing animatedly amongst themselves and mingled with their calls was the "squeaky gate" voices of several long-tailed, noisy grackles. Soon they will be seen swaggering across lawns and open fields in their quest for nourishment.

Early mornings now robins can be heard caroling, but they are not yet in full voice. The soft, sweet warbles of a bluebird are enough to bring Spring if anything could. The bluebird's voice seems to coax the sun to shine if it isn't already and to shine warmer still if it is.

More often these days the cardinals can be heard with a hint of insistence in their voices as they prepare to establish their territories for the new season. The males place themselves conspicuously in the tops of trees or shrubs so that they can be readily seen by their rivals, while the females go about their business as though nothing out of the ordinary was going on.

The chickadees are calling “knee-deep” even though not a speck of snow can be found. They may well find themselves knee-deep before the month is over, for Old Man Winter isn’t one to give up without his usual struggle. A hoarse, chickadee-like call back in the woods announces the presence of a titmouse that also is ready for winter to be over.

Rose-washed male house finches sing joyously almost everywhere it would seem. They are similar to their cousins the purple finches that, for the most part, prefer the more northern conifer forests, in which to raise their families. A countryman remembers their beautiful songs in the early spring days when he lived in Maine. They visited his feeders and even nested nearby and could be heard singing sweetly well into the summer. They are a deeper raspberry color and their voices are sweeter than those of the house finches. The house finches were brought to New York City from the west coast many years ago to be sold as cage birds, but then laws were made prohibiting this commerce and they were ultimately released and have been spreading westward ever since.

The melancholy cooing of a mourning dove was heard one morning and when a man hears his first killdeer crying piteously over a pastureland he will know that Spring has finally arrived even though there may be several inches of wet snow blanketing the countryside. Until then he will listen, watch and wait as one by one the many voices of springtime mingle in joyous utterance across the awakening land.

The winds of March are the winds of change, the changing of the guard so to speak as winter retreats before the slow but steady onslaught of another season moving inexorably northward to take its place.`

Squall lines of snow move briefly across the ridges and out across the fields followed by periods of sunshine. Clouds scud across the sky like spring lambs capering over an azure pastureland.

Some mornings are bright and filled with warmth and bird-song while others are cloud-filled, the remnants of those that left a thin blanket of snow on the ground overnight. On one such morning a man took a short trek about the countryside to see who had been out and about since the snowfall and before the sun made the tracks indistinguishable.

Crow tracks showed every detail of the bird's foot. Each crease and line, like a fingerprint, was clearly delineated. Rabbit tracks reached out from brush-heap to vineyard and back to woodlot and hedgerow. Along one such set of tracks a farm cat had sat for some time surveying the area. Its heel and footprints as well as a carefully outlined sitz-mark were imprinted in the snow.

Farther along at the edge of the cemetery, a chipmunk crossed the roadway leaving its smaller version of squirrel tracks to mark its passing. Near by a gray squirrel had left its larger signature.

The vernal equinox is but a few days away and back in a brushlot a song sparrow practices its spring song. A man likes to hear the first cautious notes of a song sparrow at the tail end of winter because he is confident that warm weather is on its way. This isn't to say that this will be the end of bluster and snow, for many a year a man has watched a song sparrow sing his cheerful song as heavy, wet snow fell all about him and flecked his striped brown coat with white. The harder it snowed the louder the little brown bird sang. There is a lesson there for all of us.

Each day now a flock or two of geese can be seen or heard as they pass overhead. Some flocks are actually heading in a northward direction and soon those from the southland will be filling the sky with their V-formations and their joyous honking. Their inner clock is telling them that it is time. Little by little the land and its creatures are awakening and soon the great primal pulse will quicken with new vigor.

A man considers himself fortunate to have been outside at the right time one bright morning. He heard voices in the distance, voices that struck a chord in his memory but it was several moments before he remembered who the voices belonged to. Not discordant but to the contrary having a musical quality about them. They sounded much like "woo-woo, woo-hoo" as the huge white birds passed overheard. These were whistling or tundra swans heading for their far northern homes. It has been many years since a man last saw and heard these magnificent birds. Their long necks outstretched and seven-foot wings steadily beating, it was no time before they were lost in the distance.

A countryman is always filled with awe and wonderment whenever he is witness to any one of Nature's spectacles such as this. How empty the world would be without the wild creatures. How impoverished would be the soul of a man without the many masterpieces of Creation.

# SPRING

The vernal equinox has now taken its place in the celestial calendar. Daylight hours increase more rapidly now while the period of darkness becomes shorter. The internal clocks of all wild things have marked the time and they respond accordingly while the pulse of life quickens throughout all creation.

Cold frosty nights followed by temperatures well above freezing stimulate the sap to rise and fall up the tree trunks. Eventually it will extend to the outermost tips of the smallest branchlets.

When the sap begins to flow, a countryman remembers the years that he tapped his maples, gathered the sap in five gallon buckets and carried them to the house where his wife boiled the sap down to just the right concentration. There is a knack to this that comes only with experience.

Homemade maple syrup and sugar has a flavor that the commercial syrup-makers can never quite achieve. The secret, the man thinks, lies in boiling the sap as soon as possible after it leaves the tree and in smaller batches. There is no time for fermentation to set in as is the case where the sap is held for a period of time in bulk containers.

One would never become rich by this method but the riches obtained are the clean, true, full-bodied flavor of the maple in its purest essence. What a pity that so few people have ever realized the true substance of the maple. Those who have know the pure delight of syrup on pancakes or hot biscuits. After the breadstuff has been swallowed, the true maple flavor lingers to be savored.

These are the thoughts and memories of Spring. Others have spring-like thoughts, too, but in a different vein.

High overhead in the clear morning sky circle two red-tailed hawks. At first glance, one would assume that they are looking for breakfast. This perhaps is true but there is more involved here than a satisfied hunger. For a time they glide and circle and hover quite some distance from each other. A man watches for some time, hoping to observe what he has in other vernal seasons, and this morning he is not disappointed.

One of the hawks rises higher and higher on broad wings until he is well above the other. Lazily he circles a few times, then, folding his wings, he drops towards the other hawk, pulling out just before impact with the other hawk. This is repeated several times and, to the uninitiated, it would seem that there was a conflict over territory or air space. All of this, however, is part of the annual courtship and bonding ritual. The repeated dives achieve varying results. Sometimes the female, the lower hawk, will dip downwards suddenly with a half-roll or on occasion she may surprise him by rising to meet him, turning deftly onto her back and clutching at him with her talons. These aerobatics are typical of the raptors.

Mourning doves and robins greet the dawn each in their own special way. A man is awakened at times to the soulful call of the dove while the robins energetically call "cheeri-up, cheeri-up". Even on an overcast morning, these voices have a tendency to brighten the early hours.

No matter what the weather may be, spring has arrived, and right on time just as it has for many millennia.

April showers are not always rain showers. Sometimes they are snow showers, wet sticky flakes that adhere to tree trunks, fence posts and your coat. The wet flakes form bonnets for the crocuses and bow the heads of golden daffodils. Then the clouds move apart and let the warming rays of the sun shine through and soon the snow is sliding off the fenceposts and trickling down the roughly grooved bark of the old maple trees, soaking into the already damp earth below and providing moisture for the sap that is already rising to nourish the new growth on the outermost branches of the old maples and all the other trees.

Rain, snow, mist and fog all contribute their moisture to sustain the lives of all living things, and April is the month of moisture.

The rain of the night stopped during the wee small hours of the morning and before long the incessant wind had dried the surface moisture. The temperature dropped below freezing and the puddles skimmed over with ice. Shortly after dawn, the voices of robins caroling could be heard and a cardinal called cheerily from a brush lot.

A lone red-winged blackbird is perched at the top of a sapling and calls "chek-chek-chek" as a man approaches. Drawing closer elicits a more aggressive response from the red-wing. Now he gurgles "o-ka-lee" and spreads his tail feathers for emphasis each time. At first there was only a narrow yellow bar showing on his shoulder but as a man approached closer still the red-wing became even more assertive of his territory and as he continued to call his crimson epaulets began to show as he slightly opened his wings each time. A countryman respects the defense of one's territory whatever the creature may be and so he slowly withdrew, allowing Mr. Red-Wing to reign supreme over his meadow domain, at least for now until another red-wing male came along and challenged him.

The starlings have exchanged their heavily spotted winter coats for showy iridescent spring suits with only a few speckles showing through, and also now sporting bright yellow bills.

An introduced species from Europe, starlings have displaced many of our native species such as the larger woodpeckers and bluebirds. These are all cavity dwellers and so is the starling. Being very aggressive, the starling usually evicts the rightful owner of a nesting site and appropriates it for itself, leaving the original occupant to find or make new quarters. This may upset the breeding cycle and result in the loss of a brood that year. This is one of the contributing factors in our loss of certain species, or, at best, a decided decline in their numbers.

The starling is not all bad as it devours great quantities of noxious grubs and beetles as well as other pests. The problem with any introduced species is that they often have no natural enemies in their new environment and their numbers increase more rapidly than the native species, thus contributing to their decline.

A countryman knows that Spring is here in spite of the frequent inclement weather for just last week he saw a killdeer patrolling its nesting grounds and plaintively crying "kill-dee, kill-dee" as it scurried over the rough ground on its long, slender legs.

With a few warm days for encouragement, tree buds swell and show pale shades of color of the new life hidden beneath the protective sheaths. Catkins commence to show on the quaking aspens and pussy-willows scamper all over their natal branches.

Newly sprouted grass colors stream-edges a vibrant green and provides new greens for the herbivores such as deer, rabbits and woodchucks.

Speaking of woodchucks, they are up and about these days and one ambles across a pathway in front of a countryman, pausing to ascertain just what he is before continuing on its way.

The warmer temperatures have thawed out the ground and earthworms are seen crawling about aimlessly, much to the delight of the robins which have been seen eating the tart fruit of the sumacs. Robins and earthworms seem to naturally go together much as woodcock and earthworms do. The chunky little brown birds with the long flexible bills should be arriving from their winter homes most any time now.

Flickers call loudly back and forth and a hairy woodpecker drums on an electric pole. He drums for some time, apparently pleased with its resonance. Nuthatches hitch their way up and down nearby trees.

As darkness covers the land with a velvet cloak, spring peepers call clamorously from a distant marsh, having only recently awakened from their slumber beneath the marshland muck. The warm spell was their alarm clock, telling them that it was time to arouse from their dormancy and consummate their conjugal obligations for the new year. Considering the capricious nature of the season, they may be sent back to their bedchambers several times due to freezing temperatures. Cold-blooded creatures that they are, they depend on outside temperatures to activate them. The warmer it becomes the more active they can be.

A few days later and an early spring snow storm does send the peepers back to bed. The cold stops the flow of sap in the sugar maples and sends most of the wild creatures scurrying for shelter.

All day the storm raged over the land and into the evening, but by dawn the next day it had blown itself seaward and the sun commenced its work of melting the accumulated white crystals.

By the following day the world was back to normal and the temperature was well above the freezing mark. Song sparrows sang joyously once more from every brushy tangle, or so it seemed. White-throated sparrows called from their concealments in the hedges and thickets while the red-wings and cardinals attempted to outdo each other with their vocalizations. After all, they must get back to the unfinished business of establishing territories.

The day ends as the one before it with a pastel-shaded sunset and relatively clear skies. The air is still mild and a lone peeper calls from a wet woodland. Later on, as the darkness deepens, it is joined by the peeper chorus from the distant marsh. Although heavy patches of snow linger here and there on the landscape, things are back to normal.

The clear sky allows a man to see the comet above the northwestern horizon. Hale-Bopp glows in the evening sky, its broad tail resplendent against the deep darkness of outer space.

The old adage that the peepers have to freeze up at least three times before it can really be spring holds true most years and it certainly holds true this year. The little Hylas seem to be out and in concert for a few days when a cold spell arrives and they have to pull their brown muddy coverlets over their heads for warmth and protection.

These diminutive tree frogs are hardy little creatures. If the air warms sufficiently and maintains a temperature in the upper thirties or so they will be back out and tuning up even though the edges of their pond or marsh are covered with snow. When the temperature reaches towards the sixties it is not unusual to hear the peeper chorus all day long, but it is in the evenings when human activity and its related noise has abated that the orchestra can be heard at its best.

Although Old Man Winter hates to leave, yet signs of spring continue to increase. The grass is becoming greener and the thin layers of snow that come and go seem to increase its color.

Leaf buds are opening on the bramble canes and the new green leaflets peek out cautiously at the slowly warming world. They are content to wait though until milder days arrive to open completely and the elderberry bushes are following their lead.

Some small bushes are more hardy and have already unfolded their small green leaves covering their once bare branches. Catnip plants have put forth new aromatic leaves on short stems, keeping a low profile for a few more weeks yet all the while ready at any moment to burst forth with rapid growth when warmer days arrive.

The male flickers consider that it is time to start serious courtship. They drum loudly on dry dead stubs, dead limbs and utility poles. Sometimes a metal downspout seems to suit the taste of an ardent suitor. Anything that will make a resonant sound to let the females know that they are available. When one or more discover a lady flicker in the area they surround her, bowing courteously up and down before her and fan their tails as they ecstatically call "wick-up, wick-up, wick—up" a half-dozen times or more. Each male does his best to make a good impression and it is sometimes quite comical to watch as each one tries to outdo the other one. Sometimes the female responds by tapping a few times on the limb she is perched on.

Despite recurring cold blasts of wind and snow, the various birds are claiming their territories and house finches will attempt to beat the robins at nest-building. Woodchucks have finished their spring house-cleaning, their burrows showing fresh earth and debris at their entrances. Small insects dance in the sunlight on a warm day and a countryman listens each day now to hear his first phoebe. That first call may well be heard on a bright morning when frost covers the roofs but the tiny ice crystals will not deter our earliest flycatcher from arriving on time.

The sights and sounds of Spring are becoming more numerous each day and a countryman finds more of interest as April edges ever closer to May. Cloudy and misty, rainy and even snow-showery days remind him, however, just how fickle the month can be some years. There are the bright and seasonably warm days that are prone to give one a touch of Spring fever.

On one of those lovely mornings a countryman discovered a host of sights and sounds of Spring. He even discovered something of the essence of early last Autumn. Low to the ground, fastened securely to a sturdy stem of dried grass, was a tan-colored egg-mass laid there by a female praying mantis before the killing frosts ended her life cycle. When the weather becomes sufficiently warm, the eggs will hatch into thousands of miniature mantises which will immediately go on the prowl for insect prey.

Two male house finches argue over a female on the lower branches of a sapling while, perched on the top-most branch, a song sparrow proclaims the area as his own, unmindful of the twittering going on below.

Field sparrows call back and forth across the vineyard rows, their voices adding to the pleasantness of the day, while in a distant hedgerow the slurring calls of a red-bellied woodpecker can be heard and some other member of the woodpecker family drums out a merry tattoo.

In a distant field several wild turkeys busily engage themselves in the serious business of their spring mating rituals. The hens feed nonchalantly while several toms fan their tails majestically and strut about impressively. They are fascinating to watch and sometimes comical as well, when one turkey tom tries his best to outdo another one in display.

A trim little titmouse flits about some bushes, its large dark shoebutton eyes watching a man as it looks for something to eat. Its cheerful "peter, peter, peter" call can be heard as it disappears into the nearby thin woods.

Not to be outdone by its house finch neighbors, a female robin busily builds her nest in a rhododendron bush where she has raised many broods over the past several years.

As the day wears on, clouds move across the sky and thicken as evening approaches. A robin chirrups a cheery farewell to the passing day while dusk deepens and the blanket of night envelopes the countryside.

In the distance peepers can be heard calling as a light rain dimples the surface of their pond and newly wet ground emits the pleasant and earthy smell of Spring.

Snowflakes fell, slanting with the raw wind that blew out of the northwest, from the leaden sky overhead. It was a morning when the mercury in the thermometer had risen just above freezing and a countryman did not expect to be out for long as it was doubtful that many creatures would be braving the elements. But he was wrong. They had things to do and they were not about to let the elements interfere all that much.

Walking to a more sheltered spot at the edge of a fallow field near a woodlot, he paused to watch the white crystals as they fell through the air. He paused to watch and to listen and mostly to listen for he was more apt to hear than to see on this cold and windy morning.

First there was only the distant cawing of the crows to be heard. Then, close at hand, the silence was broken by the bubbly "teakettle" song of a song sparrow and from some hidden spot at the woods edge came the "sree" call of a chewink or towhee. In some parts of the country they are called "ground robin" referring to their chestnut-colored sides and dark backs and their habit of scratching noisily in the leaves for bugs and worms.

As a man waited and listened for the towhee, several goldfinches alighted in a sapling close by. He noted that they were still in their winter plumage but in a few more weeks they will be attired in their summer suits of gold.

The melodious trill of a ruby-crowned kinglet came from a dense thicket and a chickadee called "knee-deep" several times as it passed by.

A cowbird uttered its creaky call and spread its wings slightly and bowed from its perch as it did so. Patrolling its territory, a red-wing alternately called "cheer" and "chek", both call notes used throughout the breeding season, as well as the gurgling call for which they are noted.

House finches vigorously called from all over the countryside. White-throated sparrows called, for Old Sam Peabody but he declined to answer their summons so they continued their sweet calling. From somewhere out in the field, a field sparrow sang its plaintive song.

A countryman was surprised to hear so many of his neighbors this late April morning. Over the years he had become acquainted with all of these and many more and was able to match voice with owner. The more time one is able to spend in the out-of-doors the more familiar with its denizens one will become. Watch and listen as they each contribute their individual part to the whole scheme of things.

The snowflakes ceased falling and the sun glowed weakly through the thinning clouds. As the sun shone with more intensity the air warmed quickly several degrees and the first tree swallows coursed back and forth across the clearing sky. Spring was here to stay; all Nature proclaimed it.

With the arrival of May the regenerating of Spring is working rapidly towards its climax. Leaf buds that were just starting to open a week ago have now burst into full leaf. These are the early bloomers, so to speak, but the rest will follow in quick succession according to their biological clock and increasingly warm temperatures.

Catkins dangle loosely from twig tips and the spring flowers push eagerly upwards out of the soil. Their leaves unfold, stems grow upwards and they will blossom out at the right signals given by sun and rain. The earliest of the flowers are already in bloom. Some have already bloomed such as the hepaticas and arbutus.

Late April and onwards is the season that a countryman likes to call the golden season. It usually starts with the dandelions and when they bloom the goldfinch males have changed into their golden summer suits complete with black sleeves and tails. About the same time the yellow-flowered coltsfoot brightens wet or damp areas along roadsides and elsewhere. Yellow mustard buds unfold and soon whole fields are aglow.

Bumblebees are busy and hornets already are stripping wood pulp from dead or dried wood to start their new paper houses.

Woodchucks munch hungrily on tender new greens as do also the cottontails who now are in an amorous mood and can often be seen by twos and keeping close company. Mourning doves and killdeers have also paired off while a man knows of at least two robin's nests that have five eggs each in them.

Chipping sparrows greet each morning with their lively chipping song, adding their voices to those of the early-rising robins that greet the dawn, their carols ascending well before full light.

Meadowlarks are more abundant this year and a man hears them every morning calling "to see you, to see yer". They can be seen in close-cut fields and also dropping into or arising from fallow fields and meadows.

Sparrow hawks or kestrels perch on utility wires and scan the ground below for prey. The little falcons look for mice and voles and an occasional small bird. Later in the season, they will harvest great numbers of grasshoppers offsetting any perceived loss in the songbird population.

The warmer days and nights have aroused the toads from their beds. From farm ponds and marshlands their trills can be heard joining with the shrill calls of the hylas or peepers and contribute to the sounds of a spring evening.

Leaf-buds and flower buds unfurl and unfold revealing the new life that has lain dormant through the winter.

Silver maples display yellow-green leaves that hang from the branches like so many tiny umbrellas. Tiny maple flowers are also in evidence, and these will later produce thousands of winged seeds that will cover the ground, porches, sidewalks and twirl away with every errant breeze. The native maples soon will repeat this process.

Chokecherry tree branches are loaded with fragrant blossoms, attracting bees and other insects which in turn will attract newly-arrived birds that will need food to recuperate from their long journey north.

Almost every day now a different species arrives in the area. Catbirds have arrived and mew from the under growth or can be seen eating the wine-colored fruits of the sumacs.

In the distance a cock pheasant calls "awk-awk", telling the world that he has laid claim to that particular part of the countryside.

While the white-throated sparrows have pretty much moved on towards their northern breeding grounds, their cousins, the white-crowned sparrows have temporarily replaced them. Their soft sweet calls can often be heard even on a chilly morning as they rummage about beneath the bushes much as the white-throats do.

For several days scattered small flocks of cedar waxwings were in the area, calling softly to one another. One flock was, you guessed it, in a tall cedar tree.

On one of the rare balmy, sun-filled days, a turkey vulture was seen floating lazily across the azure sky. Lower down, over a meadow, a harrier or marsh hawk searched intently for meadow voles.

Barn swallows have now joined the tree swallows as they sweep through the air, twisting and turning, dipping then rising abruptly as they pursue their insect prey.

The phoebes have finally arrived and their cheery notes ring out on the morning air about the farm buildings. Other members of the flycatcher family will be arriving most any day now. The crested flycatcher and the pewee are residents while other smaller members pass through on their way to the north country habitats of spruce bogs and where the deciduous forests merge with the northern conifers.

A northern yellowthroat called tentatively one morning from dense bushes. Apparently it was a transient on its way further north as it has not been heard since. The resident males will be here soon and their lively calls will be heard well into the summer.

The adderstongues are in full bloom, their yellow, down-turned blossoms nodding in the almost incessant breeze. Their splotched olive and green leaves are distinctive and bestow upon them their other name of trout lily.

Everywhere there is renewal and regeneration taking place once more. Natural and man-made scars upon the earth are being healed and covered over with grasses, plants, shrubs and vines. New life and new growth are seen everywhere, all signs of the continuing work of the Creator.

Golden-maned dandelions prowl up and down the vineyard rows, rubbing shoulders with the wild mustard that soon will outstrip the "lions" in height. All across the lawns they also roam, their golden manes shaking in the springtime wind.

Wild strawberry plants are conspicuous with their showy blossoms, indicators of shortcake in June. A countryman makes a mental note of where the blossoms are most prolific and will return, dish in hand, as he has for several seasons now.

The bubbling voices of bobolinks can be heard coming from hayfields and meadows. Nests built in hayfields are at risk during haying time, but those hidden away in the meadowlands will assure another generation of bobolinks will survive.

Wild turkeys can be seen feeding in fields and along woods-edges. May is the month for tom turkeys to strut and make huge, splendid fans of their tails and gobble boisterously. All of this display is done to acquire a harem of hens and provide for the continuation of the species. A man looks forward each spring to hear and see the pheasant and the turkey and the partridge as they perform their courtship rituals.

Small purple violets add color along one of the numerous woodchuck trails that can now be seen radiating from burrows out into fields and vineyard headlands. As a man ambles along, one of the sturdy whistle-pigs sits up, ramrod straight, its front feet dangling over its chest before it dashes to the safety of its den. Its shrill alarm whistle can be heard as it disappears from sight.

A tall cherry tree, loaded with creamy blossoms, attracts four beautiful male orioles in their bright orange and black suits. They call out cheerily as they forage for insects. This day they feed together amicably, with no apparent thought about territory or a mate. It seems to be an avian bachelor party. It won't always be like this. Any day now, their bright colors will be seen flashing overhead as they chase each other here and there in the competition for nesting rights.

In contrast, two colorful male goldfinches flutter, beak to beak, several dozen feet over the vineyard, apparently in some disagreement.

May is warbler month and the first ones to arrive are the myrtle, or yellow-rumped warbler and the yellow warbler. A yellow-rumped warbler is working the spruce grove, quietly inspecting each branch. Several yellow warblers are going over the branches of some willow bushes. From a briar patch comes the rollicking "witchity, witchity, witch" of a northern yellowthroat.

Soon their voices will be heard from tangles and thickets all over the countryside. A man likes to think of them as little highwaymen, as the males peek out from some thicket their black masks contrasting with the bright lemon-yellow of their throats. They keep a man company when he picks the red strawberries in June, and they continue their cordiality as he harvests the various bramble-fruits in July.

A countryman enjoys the many colorful warblers and he, like others before him have, consider them the butterflies of the bird world.

The voices and colors of spring become more abundant and varied as the season unfolds towards summer. By the end of May many warblers and numerous other species can be seen and heard. Some of the birds pause briefly on their way farther north to rest and feed while others settle down if they find suitable habitat or if they have nested here in years previous. Food, of course, and water are also considerations.

A countryman listens to the coarse calls of a male red-bellied woodpecker as it pokes into and probes about the dead top of an ancient ash tree. In the distance, its calls are answered by another woodpecker.

A brown thrasher vocalizes from a branch overhanging the edge of a fallow field. Dressed in a rich, rusty-red suit with a heavily streaked vest, the thrasher surveys the countryside through bright yellow eyes. A man is always amazed at the many sounds and notes that issue from the throat of the thrasher. The mockingbird, thrasher and the catbird are sometimes classed together as mimic-thrushes although they are not true thrushes. They have long tails, often expressive with movement, strong legs and slightly decurved bills.

A rufous-sided towhee calls "drink your tea" and intersperses that admonition with an emphatic "sreeee!", all the while remaining concealed in the undergrowth. Towhees are more often heard than seen.

An explosive "chebec!" betrays the presence of a least flycatcher in the area. Some minutes pass before a man spies it as it darts out from its perch in pursuit of a flying insect.

Two hummingbirds zip over and around the bushes and shrubs like miniature supersonic jet planes. A dark blue indigo bunting watches the goings-on from its perch in a scraggly sapling. Soon its song will be heard from the woods-edges as summer's heat starts to build. For now it quietly watches its neighbors and all their many activities.

A flash of flame-orange and black in a willow bush attracts the attention of a countryman. Focusing his binoculars, he finds the form of a redstart, one of the colorful wood warblers.

As the day wears on it becomes even more colorful, especially in one area of about fifty feet or so. A bluebird sits facing the sun, the warm rays striking its russet-colored breast. The sun turns a cardinal a crimson hue as it whistles cheerily close by in a bush. Gray catbirds with perky black caps play hide-and-seek in the undergrowth, their mewing cries disclosing their whereabouts. Ebony redwings call and flash their epaulets. Yellow warblers and male goldfinches add a touch of feathered sunlight to nature's palette.

Afternoon shadows lengthen and deepen as the sun sinks slowly out of sight. A wood thrush calls melodiously from the dusk-shrouded woods, its flute-like voice a benediction to the declining day.

Clouds steal across the darkened sky and toads trill harmoniously from their seats at the edge of the farm pond. They pay no heed to the light rain that falls gently over them and a pale moon peeks through a rift in the gathering clouds.

The early June countryside is slowly edging towards summer. Baby robins hang their chins over their mud and grass-lined cradles, their mouths gaping eagerly when one or the other of their parents arrive with food.

The waves of warblers were not as heavy as other years. Perhaps this was due in part to the abnormally cold spring and increasing destruction of South American rainforests no doubt contributes to a decreasing population.

Also this year they seemed to be more quiet and more difficult-to locate by voice. Nevertheless, a countryman was able to see chestnut-sided, blackpoll and black-throated blue warblers before the migration tapered off.

A rose-breasted grosbeak calls from a tree-top, its mellow, robin-like voice ringing out across the countryside. Somewhere, back in the underbrush, the monotonous call of "cucucu, cucucu, cucucu" issues forth, telling a countryman that a black-billed cuckoo is in the area. Later on, he sees the bird, perched low down in a sapling, a characteristic of the species. Unlike their parasitic European relatives, the black-billed and the yellow-billed cuckoos build nests and raise their own young. They also are some of the few birds that will eagerly eat the destructive and hairy tent caterpillars. They seem to arrive about the same time that the moths have laid their eggs and the small larvae weave their silken tents that they will call home for weeks to come or the food supply runs out.

Wild phlox is commencing to bloom, their white and rose-pink flowers sometimes filling the air with their sweet scent.

Late spring evenings are a period of the day when a man likes to go for a stroll and listen to the wood thrushes call back and forth from the slowly darkening woods. In more northern climes the wood thrush is joined by the hermit thrush and the veery, another of the musical thrush family. No matter the location, the melodic voices of the thrushes are pleasing to the ear.

Years when there is an unusually cool spring, the evenings cool down dramatically and this temperature change affects the cold-blooded creatures such as the toads and peepers.

Their activity decreases with the temperature and by nine or ten o'clock at night the voices from the farm pond are slow and laborious. Two' peepers call slowly "pee-eeep, pee-eeep" back and forth while a pale half-moon smiles down at them from the clear, star-studded sky.

The early days of June pass, full of promises, promises not only of new life but of ripenings and maturings and fulfillments.

Dandelions, which only a short time ago were shaking their shaggy yellow manes, have matured and their seeds parachute away on the wind to start a populace in some distant region. Of course not every seed finds a fertile resting place, contrary to the thinking of those who desire and cherish dandelion-free lawns and landscapes.

Chokecherry blossoms have covered the ground beneath the trees with their snow-like petals, leaving behind tiny embryonic green cherries which will, in time, develop into mouth-puckering fruit which only the birds and other wild creatures will truly enjoy. A countryman always will try a few of the fruits in season, if only to refresh his memory of the first time he tried them at a much earlier stage in his life.

In contrast, the black cherry trees are late-bloomers as their flower buds are just nicely opening.

Woodchucks are commonly seen as the vegetation becomes increasingly lush and is still tender and succulent. As the weeks go by they will be seen less as the grasses and weeds increase in height and provide heavier cover and concealment for all the wild creatures.

In the shadows of the little ravine a striped chipmunk pauses a few brief moments to adjust the contents of his pouches before disappearing into a tangle of fallen branches and ground litter.

Flower buds on the bramble canes are swelling, the most noticeable being those of the blackberries although the black raspberries will bear fruit several weeks before the old fashioned blackberries will. Perhaps it is because they are a larger fruit and also have more prominent blooms on much larger canes.

Wild strawberries will be setting fruit in a few weeks, that is if there have been sufficient bees to pollinate the blossoms that are so abundant. Then too, the cool, wet spring has been a deterrent to the activities of many of the pollinators.

Goldfinches dip and rise in their flight over the vineyard while an indigo bunting makes melody from its perch along a headland. A cottontail chews clover in the shadows beneath where the bunting sings. As a man approaches, the rabbit turns and disappears into the nearby weeds.

A robin with a beak-full of wet mud and grasses pauses a few moments before continuing to its construction site. Other robins in the area have young ready to fledge. It's all a matter of timing.

Gray tree frogs give their bird-like calls as clouds build across the afternoon sky and continue on into the evening when they are joined by toads trilling in the now gently falling rain. A couple of peepers provide soprano parts to this late spring amphibian chorus.

Meanwhile, life stirs actively in the gelatin-like masses of frog and toad eggs that were laid in recent weeks. Soon the ponds and pools will be alive with wriggling tadpoles, frogs and toads-to-be.

For now, the buds and eggs and unhatched pupae are promises of a new generation that will continue to provide the various life-forms that are so familiar to us and are a necessary part of our environment.

A man ponders what the future holds and what those future seasons will be like if the ponds and pools and forests continue to be contaminated by chemicals and chemical residues. We have much to lose and much to gain, depending on the choices we make in the future.

Just as a countryman has gotten used to spring it seems to slide right into summer. Showery days followed by a few hot ones combined with mild nights, and flower buds swell and burst into bloom.

Flower buds such as are found on the blackberry canes open and the canes are all aglow with the white blossoms. Some blossoms are still imprisoned within their swollen green buds but soon they will join the others in their march along the thorny branchlets.

A song sparrow sits sideways on a leaning bramble-cane and pours out a melody that is as much summer as it is yet spring.

Black cherry trees are resplendent with creamy-white blossoms that hang in racemes four to six inches long from the branches. Their scent mingles with that of a nearby patch of phlox.

The sweet smells of spring are everywhere; the scent of new grass, new leaves, all the varied kinds of blossoms as well as the ripening of certain grasses that do not wait for summer to officially arrive. Already, small green fruits have replaced the blossoms of a few weeks ago.

Swallows and chimney swifts are often silhouetted against the azure sky in their quest for insect fare, not only for themselves but for their hungry nestlings.

Young robins are seen hopping about the door yards, their spotted vests indicative of their membership in the thrush family. They noisily beg to be fed by their parents who are already seriously considering the idea of a second brood and are making the fledglings look for some of their own food in between feedings.

A yellow-bellied flycatcher darts out now and then from willow brush perches to catch and consume some of the numerous insects that are available at this time of the year. Although it is another member of the look-alike Empidonax flycatcher group, its spiritless and simple "chu-wee" call assures a countryman that it is indeed a yellow-bellied flycatcher,

Low, flat buzzing calls from the same low-lying, scrub-filled area intrigue a countryman. Rather expecting to find a golden-winged warbler, he is surprised to discover that the calls are made by a clay-colored sparrow, a bird that he is not familiar with and is another addition to his life-list of birds.

Hot days followed by warm evenings are just what the gray tree frogs have been waiting all spring for. As a man watches the sun set from a favorite vantage point, the tree frogs blend their voices into a veritable chorus, providing a splendid benediction to the day's activities.

# SUMMER

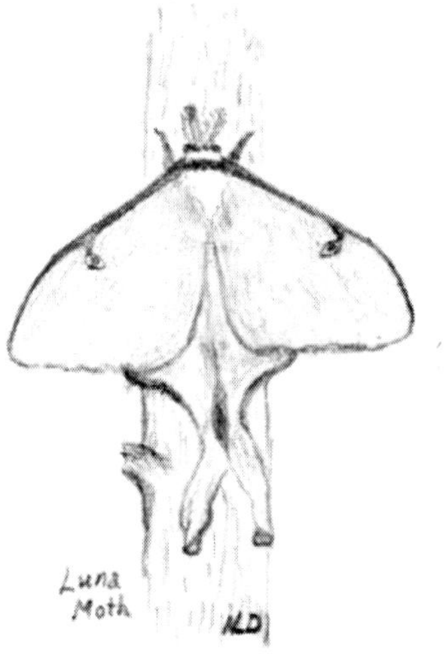


The Summer Solstice unobtrusively slips Spring into Summer. It is a quiet process, no fanfare or jubilant celebration among the wild folk. Quite the contrary, the migrations are over and territories have been established. For some of the birds, one brood has already fledged while for others there are young still in the nest. Animal babies are seen close to home under the watchful eye of Mother or else have only recently struck out on their own.

Young gray squirrels, scarcely larger than their cousins the red squirrels, are seen munching on maple-seed salad, an item that has matured just in time to furnish food for the young squirrels.

Young cottontails and woodchucks are learning the rudiments of survival in a not-always kind world. Their mothers teach and guide and discipline, but ultimately they must survive by being alert and constantly aware of their surroundings, aware to what the various voices and noises mean in the world around them. Heedlessness on the part of a young woodchuck was observed late one afternoon. The youngster apparently was exploring the area around his home or feeding on tender greenery in the tall grass when a farm dog came trotting along the nearby vineyard road. Unaware of its imminent death the little woodchuck was soon dangling from the dog's jaws and heading for parts unknown. One moment it was enjoying life, the next its brief life was over. Survival is for the observant.

The catbirds mew less frequently now that youngsters occupy their time and energies. So it is with most of the other birds. Wrens sing their bubbling little arias as they busily provide for their families and the clay-colored sparrow buzzes from the bushes. No doubt its mate is incubating on a nest hidden somewhere close at hand.

A man spies a kingbird perched on a fence wire. Spying an insect, it flies out in a quivering sort of way, seemingly flying "on the tips of its wings" which is characteristic of the species. Both sexes are dressed in a similar manner with dark slate-gray heads and backs and tails, the ends of their tails tipped with a broad band of white. A white vest completes their outfit resulting in a very neat and attractive-looking bird. An often concealed crimson patch on the crown of the head contributes to their name. Then too, they are aggressive and can be pugnacious when defending their territories, attacking hawks and even eagles where that species is present.

The bramble bushes are starting to set fruit while many blossoms are still in evidence. Wild strawberries are ripening right on schedule, but their flavor does not seem to be as full-bodied and sweet as it should be. Not enough bright sunlight could be a contributing factor. Flat-topped blossom bud clusters adorn the elderberry branches. Soon they will unfold into creamy blossoms.

Summer flowers are starting to bloom while others are still in bud. A man pauses to smell the delicate perfume of the wild rose whose pale pink petals are unfolding along the roadside. The bud and bloom of the wild rose is a sure sign that summer has indeed arrived.

All sorts of fruits are growing and maturing as the Fourth of July approaches. Wild strawberries continue to beckon to be picked before their time in the sun declines and it is time to harvest black raspberries. One harvest tends to merge right into the next one.

All Nature is busily producing both cultivated and wild crops. Small green cherries, raspberries, and black berries are all developing and maturing as each day brings with it its measure of moisture and sunlight and the necessary warm that generates growth.

Fields of freshly-cut alfalfa and timothy lay curing in the sun while row upon row of corn plants wave their sturdy green arms in the wind on a sultry afternoon. A gray tree frog calls languidly from somewhere in the cool shadows of the woods-edge.

The humid afternoon wears on and the sun's rays penetrate the haze and touch the thermometer causing its indicator to move ever closer towards the ninety degree mark. In the meantime, thunderheads tentatively build across the sky.

The south-west wind curls the leaves up, their lighter undersides tipping towards the sun. Often, this is an indicator of an approaching storm.

Cool thickets and shady corners conceal nests of yellow-throats and yellow warblers. They quietly go about their business of tending to their nestlings only occasionally singing against the heat.

An indigo bunting breaks the silence momentarily in the distance while several goldfinches roller-coast their way across a field seemingly unmindful of the heat waves that shimmer beneath them.

Paper wasps busily gather material from a weathered fence post and transport it to their hidden construction site. Layer upon layer is added to their oval home while inside new cells are continually being constructed to the paper comb that cradles embryonic wasps that will soon develop into adults and add their labors to the others. This expansion necessitates the removal of layer after layer of the inner paper walls, this being chewed and mixed once more with saliva to become pulp which will be added to the new construction outside. The wasps' home grows from the inside out, much as the fruit grows and expands around its core filled with seeds or perhaps a pit enclosing a single seed. Both grow to ultimately produce another generation to perpetuate the species.

Slowly the shadows lengthen even as the heat strengthens for a short time preceding sunset. Thunder mutters mildly in the distance and a man muses to himself that there just might be a thunderstorm in the offing come suppertime.

Shortly after four in the morning a robin starts caroling outside the window. A countryman does not need an alarm clock to tell him when the pre-dawn hours have arrived. However, it is much easier on the nervous system than the clatter of an alarm clock.

The first rush of nesting season has passed and more bird voices may be heard these summer mornings.

The screams of jays sound almost autumn-like at times. Perhaps it is because they have been busy with families and relatively quiet the past few weeks.

Cardinals call "whoit, whoit, purty, purty, purty" in a jovial manner. From the woodlot a flicker clearly calls "clee-up!" several times while along the roadside a goldfinch sways back and forth on a chicory stalk and says "sweetie." Perhaps he is using that endearment to call his mate.

The chicory stem the goldfinch sways back and forth on already has several blue flowers that have recently unfolded. The blossom buds open from the bottom of the plant stem and as the season matures the blue flowers bloom their way to the top. This creates a continuously flowering plant for the major portion of the summer.

Bindweed, or wild morning-glory, grows by the roadside and in places beneath vineyard rows. Its pink trumpet shaped flowers peek out at passers-by.

Creamy-white flower clusters adorn the elderberry branches, attracting numerous small insects and an occasional butterfly.

Late years, butterflies have become increasingly scarce, a cause for concern for everyone, as they are one of the indicators of whether our environment is healthy or not and also the condition of the wintering grounds of those who migrate each fall, as do the monarchs.

Recently, a man has seen a few monarchs fluttering here and there, but nowhere are they common as in past years. A black swallow-tail was seen earlier and of course the numerous cabbage butterflies. Two small yellow sulphur butterflies were observed swirling and pirouetting in the air close together. It was their mating ritual; a nuptial dance.

Two crested flycatchers were seen taking turns snatching insects out of the air. Their calls of "whee-eep!" rang out over the countryside as they continued to feast on the numerous flying insects.

These flycatchers are one of the largest in the area and nest in a tree cavity, often adorning their entrance with an old, cast-off snake skin. This may serve to deter a predator looking for a free meal.

Crested flycatchers have a bushy crest which they can erect when agitated, hence their name. They sport cinnamon wings and tail and combined with their gray breast and yellow belly they are quite handsome birds. Males and females dress similarly as do most other flycatchers.

All of this and more is part of the summer scene as a countryman rambles about the countryside on an early July morning.

Year after year as the seasons merge one into the other, a countryman has learned to look and listen for certain signs of each passing season. Plants and animals, insects and birds, all are an important part of each month in its turn. This is especially true during the warmer months of the year, yet it is applicable during cold months if only to a lesser degree or in a different form.

The pink blossoms of milkweed contribute a special yet familiar fragrance all their own. Black and red, or orange, milkweed bugs scurry about the blossoms and over and beneath the fuzzy broad leaves when disturbed. Then, of course, milkweed has its own butterfly, the familiar monarch, with its orange wings with black markings. The monarch lays its eggs on the undersides of the leaves and these hatch out into the conspicuous yellow, black and white striped caterpillars which are familiar to most country children as well as adults who go afield for one reason or another.

The crows, hawks and owls are as busy with their youngsters as the smaller, more familiar species have been. This is a critical time of the year for the young of all species. They are as vulnerable to sudden, unusual changes in the weather as they are to predators. Many young of all species never reach maturity because of these factors.

If a season is dryer or wetter, warmer or colder than is usual, then this creates an environment that can be difficult or impossible for the young to cope with.

Green frogs pluck their "banjos" from the farm ponds and swamps, while bullfrogs play their "bass fiddles" from the marshlands or lake edges.

A hunting hawk is interrupted by three crows and a red-winged blackbird as it enters their mutual territory. Dodging its attackers, the raptor quickly departs for an area where it can hunt with no interruptions.

During a stroll along a cemetery road a man noticed a mockingbird perched on a marker. Conspicuous for the past year by its absence, the mocker was indeed a pleasant surprise.

Fireflies twinkle in the gathering darkness, continuing their small contribution to the month's celebration of Independence Day.

Frogs, fireflies, milkweed blossoms and thundershowers are all part of the portrayal of mid-July. Summer has finally arrived, and as the month progresses more of its many facets will ultimately be revealed to those who take the time to wander here and there about the countryside. Each day demonstrates the creative power that is constantly at work. Each day there is something to be observed in the natural world that shows beyond the possibility of doubt that the Creator is in control and maintains and sustains all life-forms.

When July passes the mid-way mark a countryman is secure in the knowledge that the "dog days" of summer have officially arrived. There were days in June when the sun beat down mercilessly and the air was heavy with humidity, but this was only a foretaste of days to come.

The dog days are reckoned from around July 3 to August 11 when Sirius, the Dog Star, rises at the same time as the sun. After that period, when the days are uncomfortably warm and humid, a person presumably just perspires profusely, drinks ample amounts of fluids and endures the heat the best he can.

During the heat of day, most creatures wisely seek out cool, shady areas and limit their activities to that which is absolutely necessary.

A hen turkey leads her poults from the clearing where they have been foraging to the cooler climate beneath the overhanging grape vines. The poults, numbering around twelve at the time, occupy themselves investigating this new environment, while their mother watches for enemies and "put, put, puts" to them now and then and they respond in kind, keeping up a low conversation between themselves.

Later, a man discovers the hen resting in the grass between two rows of vines, only her head and part of her neck showing above the grass.

Since he had just found two breast feathers that had not been shed but had been removed in some violent manner, a countryman was concerned for her welfare. He moved slowly towards her, but after a few steps she arose from her resting spot and walked quickly over a little knoll and out of sight. Relieved that the hen was not injured, a man continued on his way.

Gray-green mullein rockets rise from their velvety launching pads, their green nose cones adorned by small yellow flowers.

Queen Anne's lace has recently opened their cup-shaped flower heads and many have reached their mature convex shape, each white blossom head decorated with a small lavender spot in the center.

The off-white blossoms of yarrow peer up through the tall grasses. A man bends to pick one of the lacey dark green leaves and crushes it beneath thumb and forefinger a bit so as to enjoy its aromatic scent.

In earlier times, the leaves were used to treat colds and as a tonic. Yarrow was one of a great many botanicals that were widely used to alleviate or even cure numerous of the ills of mankind. Now they are seldom used and few people even know about them. A very few, like the lobelia, are still a part of modern medicine, but science is supposed to produce all of our needs in the laboratories rather than going to the laboratory of nature. A countryman often ponders just where all of our so-called progress is taking us.

The sun slowly disappears, leaving only pastel shades of rose and yellow hanging above the western horizon. A field sparrow sings from atop a grape post, flying up now and then to snatch an insect out of the air.

Brown beetles buzz noisily and clumsily about the grapevines and collide with a man's head. Gray tree frogs call back and forth while high in the western sky the luminous half moon watches the activity down below.

A chipmunk "chucks" in a monotone from atop a small rockpile in the shady edge of the woods. As a countryman pauses to watch, it carefully washes its face, then behind its ears, its vest, sides and hindquarters, and finally with a flourish, it makes sure that its tail receives adequate attention as well. It sits motionless for a few more moments and then it is off, tail held at a jaunty angle. The leaves rustle as it moves through the litter on woodland floor and then all is silent save for the questioning "yank, yank" of a white-breasted nuthatch.

A countryman is out after black raspberries, but their time in the sun is about over for the year and he doesn't expect to fill his dish this day. Thunder rumbles among the ominous dark clouds which are gathering in the north. To the south, there are patches of blue sky and cottony clouds.

A green heron gives a guttural squawk as it passes overhead. It is one of the common herons but its body, tail and wings are shorter than those of its cousin, the great blue heron that also flies over on occasion.

Here and there, a few goldenrod buds are starting to open, creating veins of gold along the slender branchlets. At the same time, many of the field daisies are losing their petals and their maturing centers are turning from gold to bronze.

Silver caps the slender stems of yarrow while platinum graces the tops of Queen Anne's lace. Wherever one looks, there is treasure to be found all about the countryside as July draws to a close.

Rings of amethyst are encircling the spiny flower cones of teasel while small emeralds grace the branches of the black cherry tree.

The dark clouds move eastward, taking the thunder with them and the western sky clears considerably.

An indigo bunting sings until dusk, providing the vespers this summer evening. A lone crow wings its way across the darkening sky as tree frogs commence calling in earnest.

From the raspberry patch, a yellowthroat sings a few more times as does the song sparrow atop the brush pile, then silence reigns except for the tree frogs in the distance and the buzzing beetles.

A brilliant meteor slices across the western sky and burns out after briefly entering earth's atmosphere. Below, among tree branches, down in the grasses and through the air in between, fireflies flicker and flash like miniature meteors in the ever-deepening darkness. Behind, at each side and ahead, little lanterns light the way for a countryman as he slowly makes his way homeward.

Ripening blackberries take late July into August and into mid-summer. Gray grasshoppers rise on whirring wings from the dusty wheel tracks of the vineyard road as a countryman walks towards a favored blackberry patch. Farther ahead, the indigo bunting's song can be heard from the spruce grove or thereabouts. A man considers the bunting's song one of several that help make summer what it is. The field sparrow and the goldfinch are two more. The yellowthroat is not so vocal now, but its cheery song can still be heard now and then from some thicket or bramble patch.

Young song sparrows are trying to learn their family song but they haven't quite gotten the hang of it and the results sometimes cause a countryman to chuckle. Of course it isn't anywhere near as comical as the voices of young crows. Their somewhat nasal, whining calls radiate from the woodlots out and across the countryside, blending with the shimmering heat waves and dissipating in the distance.

A black swallowtail butterfly flutters across the field, pausing a moment at a wild carrot, or Queen Anne's lace, then down to a pink clover blossom where it spends some moments before continuing on its meandering journey. White cabbage and yellow sulphur butterflies twirl and dance in the air then flutter off among the chicory, clover and yarrow blooms.

It doesn't take a man long to harvest the few first-ripened blackberries, but there are enough to cover a morning's bowl of cereal with some to spare. A man remembers hot August days in Maine, many years past, when he often shared the huge bramble patch and equally huge blackberries with a black bear. Although never seen, the bruin left ample evidence of its foraging in and about the berry patch. Rounding tunnels and copious droppings full of berry seeds proved that one or more bears were feasting and putting on fat for winter to come. Alone, or with family, a countryman would whistle or talk so that if bruin was in the area he or she would be aware of human presence.

Bull thistles, some of them six feet tall, sport blossoms that look a lot like purple shaving brushes. When thistles come into bloom then it will only be a few more weeks and the goldfinches will once again be nesting. For now, they sing and call to one another as they perch among the grape vines or up in the big willows or dip and rise in flight over the countryside, enjoying what remains of their summer vacation.

The white blossoms of catnip attract honeybees and newly opened buds of goldenrod also bid for their attention. The busy buzzing of bee wings adds to the symphony of summer sounds.

Laying at an angle in the grass, a long, gray and white barred wing feather announces to a countryman that the hen turkey has started her summer moult while attending to her maternal duties. Crow feathers are now also seen quite often and as the birds fly overhead one often sees a wing or tail feather missing. By the middle of the month most of the smaller birds will have taken on a rather ragged appearance as they too exchange old worn feathers for fresh new ones.

Mid-summer is a time of transition, a time of maturation. It is the season to fulfill the promises of Spring that there would be a harvest. The harvest is yet in the future, but the promises are being kept in the green fruits, the green-cloaked walnuts, butternuts and hickory nuts.

The great wheel of the seasons slowly turns as summer deepens and with the deepening comes the sense of accomplishment.

A countryman feels that sunset is one of the finest times of the day. Not that he has anything against sunrises, for they also play a most important part in the scheme of things and they can be just as colorful and awe-inspiring as any sunset. Sunsets, however, mean the end of another day. Usually, the day's work is over and a person can take a few minutes to relax and contemplate the day's activities.

Walking along a familiar pathway, a man watches as the sun takes the day to another part of the world. The blazing orb reclines on a western ridge a few minutes before slipping out of sight. Then it is gone, leaving brightly colored clouds in its wake.

A man watches as the clouds slowly fade into pastel salmon and yellow shades and merge and blend into ever-softening hues.

Meanwhile, two blue herons speak hoarsely to one another as they fly to their roost for the night. They are late in returning but are in no hurry as there is sufficient light remaining.

Ripened grasses scent the cool evening air. A man breathes deeply and exhales slowly, savoring the clean sweet and unpolluted air that is an integral part of country life. Not that air is always fresh and sweet; sometimes a nearby farmer has spread a load or more of manure on one of his fields. But that too is part of country living and must be accepted as such, just as one will accept the unsavory scent of a skunk on any given evening.

Bright orange and yellow amongst the green grasses bespeaks the presence of butter-and-eggs. Once again it is that time of year for them to take their place among the goldenrod and the black-eyed-Susans whose yellow-orange blossoms seem almost to glow in the semi-darkness. The lemon yellow flowers of evening primroses open more completely in the lingering twilight.

Several bats dip, dart and swoop through the increasing darkness after their evening meal. Or, because their activity commences when ours declines, perhaps this is really their breakfast time. At any rate, they are busily engaged in acquiring sustenance after sleeping all day.

The evening star glows luminously in the western sky, long before any of the other stars are readily noticeable. This is the star, or planet, that so many people have made their wishes upon so often for countless centuries.

The thin, crescent moon stands on tip-toe its sharp lower point directed earthward. It is definitely what country folk would call a "dry" moon. Already the land is thirsty and much in need of a soaking rain, but there appears to be none of that forthcoming in the immediate future.

The first few crickets of the season chirp cheerfully, albeit slowly, as the evening air continues to chill sharply. There have been other evenings like this of late, and a man thinks that the temperature could easily be in the mid-forties once more before morning.

A countryman never knows what he will see at, or hear from, a brush pile. There is always a sense of the unexpected. Often a woodchuck will be seen scurrying to its dwelling beneath the sheltering branches. Cottontails often use a brush pile as a place of refuge in an emergency and as a winter home if a woodchuck has left it vacant. Sometimes a chipmunk can be heard chipping from its seclusion and, if a woodchuck is in residence, it may whistle sharply if one comes too close to suit it.

One day a man walked by such a debris heap and heard a familiar little voice coming from its interior. Stopping and remaining motionless, he soon observed movement amongst the branches. Soon a house wren appeared then disappeared only to reappear moments later with a companion. While watching these two, a countryman caught movement from the corner of his eye and lo and behold there was a third little brown wren. A man watched as they moved in and about the brush pile, exploring both its depths and outer fringes. Occasionally, they would chatter briefly between themselves during these explorations. A man became reasonably sure by their actions that they were siblings recently thrust out upon their own. They appeared to be having a great time and a man continued on his way, leaving them to their devices.

Although the day was warm and had a tendency to be humid, a chickadee announced that something was "knee-deep" several times. Just what that something was, a countryman had no idea, but the friendly little voice of the black-cap was a most welcome sound just as it is any time of the year.

Somewhere nearby, a white breasted nuthatch called petulantly for "Hank", but apparently received no answer. Over in the distant woodlot, a pileated woodpecker called. Its slow, accentuated vocalization rang out clearly across the intervening fields.

A countryman always thrills to hear the almost primitive calls of the pileated woodpecker and is equally excited when he happens to see one in flight or busily excavating one of its huge cavities in a tree. This crow-sized and majestic woodpecker is able to penetrate deep into the heartwood of a tree and extract a whole colony of carpenter ants, sometimes giving the tree a new lease on life. To be sure, a huge cavity remains; having a scar is preferable to losing a life. Some people erroneously think that a pileated is very destructive when they see trees with those big elongated holes in their trunks, but little do they realize what sort of insects or larvae were inside, originally, eating the life from that particular tree. The tree has nothing to lose and sometimes much to gain in longevity.

Carpenter ants and pileated woodpeckers appear often to go together. The big woodpecker also enjoys the bark beetle larvae and similar food items.

Watch for signs of this woodpecker when afield, and be assured that it and all the members of the woodpecker family are busily engaged in helping to keep our environment in balance. May we too, in turn, do our part in keeping the earth a fit place to live. This is in keeping with the original plan of the Creator.

After watching sunrises for decades, a countryman decided that it was time to watch a moonrise in its entirety. He had observed the Harvest Moon many times rising low above the eastern horizon, perhaps a half hour after moonrise, but never had been on hand to watch it as it first came into view.

This evening, only a few thick clouds stretched across the sky here and there. A luminous glow in the west was all that remained of the sunset of nearly an hour ago. Pinpoints of stars pricked through the dark canopy overhead.

Bats dipped and dodged over the vineyard and through the interlacing branches of the willow grove in pursuit of their evening meal.

From the sheltering branches of those same willows came the rasping notes of the first katydids of the season. One would call out "Katy, Katy, Katy" while another close at hand would respond that "Katy did, Katy did, Katy did". A man always wonders what it was that Katy ever did. Perhaps the intent of her supposed actions is to announce the closing chapters of summer and to give a sense of depth and feeling to the summer nights.

Both in the east and towards the west brightly gleamed a star. It was indeed appropriate that two such celestial bodies were on hand, one to bid farewell to the Day-Star and the other to greet its reflecting counterpart.

While a countryman waited in the ever-deepening darkness, crickets fiddled away in harmony with the katydids and the other members of the insect orchestra.

This was the first evening after the full of the moon and a man was assured of a resplendent moonrise. The minutes passed and by now the land was cloaked in darkness.

Then, above the eastern horizon, the gossamer clouds assumed a pastel shade of pink. The color deepened somewhat and then, there was the moon! At first it was just a thin bright edge slowly rising above the distant ridge. Fascinated, a man watched as the earth turned, revealing more and more of the bright lunar disc. Soon the upper portion of the moon was becoming veiled by the wispy clouds while the lower edge became visible just above the ridge-line.

Unlike a sunrise that cannot be watched too closely because of the bright rays, the slowly ascending moon could be carefully observed; close attention could be given to every detail of its ascent into the night sky until, at last it hung like a great golden globe in all its splendor.

Although by this time the sun had long disappeared and its western glow had faded into obscurity, its radiant reflector had taken its place. Moonlight swept across the landscape creating soft shadows here and there.

Just as the sun announces the beginning of a new day, so this night the full moon proclaimed its reign over its duly appointed portion.

In the shadows a meadow vole sat quietly, its whiskers trembling slightly and its nose quivering to catch scents of the night air. Sensing no immediate danger, it ran from the sheltering shadow into the moonlit field, disappearing into the protection of the weeds and grasses. All of the activities of the night were now in full swing.

Summer is writing its final chapters for the year. Perhaps the turkey that left a solitary footprint in the mud in passing wanted to add a "footnote" to one of these chapters.

Already, a few colorful leaf-banners are appearing on some of the sumacs. Elderberries are in varying stages of ripeness; their branches bowing beneath the weight of those which are fully ripe are begging to be harvested by passers-by. A countryman will not gather the tart fruit this year as in past years. The one who made the little pie and enjoyed its flavor is no longer here to share its tart sweetness with him.

Bumblebees hum as they grope their way over the purple burdock blossoms, then careen over to some late-blooming clover to add to their cache of nectar and pollen.

Many of the once silver yarrow blooms have turned a dull chocolate as they reach maturity. A man crushes one of the heads and enjoys the same aromatic scent that the new green leaves presented him earlier in the summer. Several varieties of goldenrod are splashing the canvas of the meadowlands with their golden hues.

Milkweed has exchanged its lavender blossoms for dull green seed pods that grow a little fuller with ripening seeds and their silken parachutes each passing day. The once dark green leaves are becoming paler as the chlorophyll departs and some plants exhibit leaves that now have hues of yellow and tan with only a touch of yellow-green here and there.

The purple bands that once encircled the teasel heads have disappeared, leaving them dark tan and even more prickly than before.

A countryman has picked the last of the blackberries for the year. The last few dishes were harvested to the accompaniment of the little cricket fiddlers. Orb-weaver spiders have grown large enough to be noticeable, their large, many-spoked web wheels seen suspended between blackberry canes and goldenrod stalks.

A few butterflies wander over the late summer flowers, stopping now and then to sample their sweetness. A man was very pleased to see a white admiral with dark wings banded near the center with white. The wing edges were bordered with small blue patches garnished with red spots.

A country-person never needs a calendar to tell the seasons. Nature provides her own calendar and marks off the passing months in conspicuous ways. Strawberries close out June, while raspberries bring in July. By the time blackberries are through, crickets are chirping and katydids are rasping away the night; it is nearly September. October is out-flanking September when grapes become succulently sweet on their vines, and when the trees become bare and wind-whipped and drenched with rain one knows that November is in full swing and that snow will soon replace frost on the fence posts and roofs. By then December has arrived, the ground has frozen down several inches, snow usually covers the ground and the Winter Solstice comes and goes. The cold strengthens, the snow increases and a new year has commenced.

As the sun bids farewell to this late summer day a wood thrush calls “pep-pep-pep” from someplace in the now-darkened woods. In a few more weeks this voice will no longer be heard. When once again those pure and silvery notes of “ee-o-lay” are heard, a countryman knows that by Nature’s calendar Mistress May has already arrived.

September seems to be a season unto itself. Indeed, it has a touch of all the seasons except winter and on a frosty morning in late September one will look twice to see that the white crystals are not snow but only particles scattered by Jack Frost.

In spring the red-tailed hawks soar together in courtship. Now the hawks spiral in the sky, sometimes accompanied by a juvenile or two, preparing for their fall migration. Those who nested farther north are usually the ones that spend the winter here.

In summer there are numerous fruits that ripen as the weeks go by. Now it is the walnuts, the hickory, beech and butternuts that are reaching towards maturity. The squirrels and chipmunks know it as well and are waiting in eager anticipation for the time to start their harvesting. One must not forget the acorns from the several species of oak. Acorns from the white oak are most favored by many species of wildlife.

In autumn, wildflowers still bloom profusely until the killing frosts cut short their tenacious lives. There are warm days and frosty nights followed by cool, crisp days. So too is September, with days of summer interspersed with cool days and crisp nights with a touch of frost in the air at times. The ninth month has a little of all three seasons encapsulated within it. On the mowings there is fresh new growth of grass, clean and green as when Mistress Spring first arrived. There is the ardor and energy of summer, as the maturity of all things reaches for its climax. A hint of autumn is already in the air as cooler winds blow from out the northwest and jays scream as they fly from brushlot to woodlot and to the hedgerows and back again. They seem to have become energized by the change in the atmosphere.

Bright colors already are showing here and there among the sumacs and the sugar maples. Willow leaves are tinged with a lemon hue.

Jewelweed blossoms yield to fat, green seed capsules that virtually explode at the merest touch, sending their seeds in all directions.

A green heron stalks about the edge of a farm pond in search of an unwary frog or two. Soon it will head south along with its relatives, the great blue heron, the egret, the bittern and others.

Swallows of all species gather in great flocks in the marshlands where there is still an abundance of insects to be caught. Family groups of swallows cluster on the electric wires between farm buildings and along roadways. Soon they will join those at the marshes and they all will move southward ahead of the frosts.

A countryman thinks of September as the swan-song of summer. In four short weeks he will watch as the final chapters of the season are written across the pages of the land.

As September reaches its median point, the local geese move restlessly up and down between the lakes, honking noisily among themselves. There are no huge flocks, just small family groups of twenty or thirty geese. Summer is nearly over; they can feel it in the air as much as a man can feel it, but in a different way. They can feel it in the changing winds and in the air currents as they fly from place to place.

Fully fledged, yet still speckle-breasted, a young robin feasts on ripe black cherries. They are dead ripe now and many are drying up on their stems. The robins and starlings and others don't mind as they gorge themselves on the dark purple fruit.

As catbirds are wont to do at this time of year, one mews monotonously from its hiding place in a thicket. It too, will enjoy the ripening small fruits for several weeks to come.

No longer is heard the "witchity, witchity, witch" chant of the yellowthroat from the willow thickets; no longer is heard the vesper voice of the field sparrow from the vineyard rows or the sweet song of the indigo bunting from along the headlands and the woods-edge.

Many summer voices have fallen silent as they quietly go about their business of exchanging worn out summer suits for fresh fall ones. Others have already started on their southward journey. Goldfinches are busily rearing their youngsters to adulthood while thistle seeds are abundant.

Grapes are ripening more each day; some of the wine grapes are ready to harvest while the Concords need a few more weeks. Then the aroma of their ripened fruit will hang cloyingly in the autumn air.

Certain members of the insect chorus can be heard most of the time, day or night. As evening wraps its velvet cloak about the land the katydids and the crickets add their instrumentals to the chorus.

Bats, large and small, twist and turn in the night air. They seem to have a certain pattern in their flight as they pursue insects out over the vineyard, through the hedgerow, up over the brushlot and back again. These patterns no doubt outline territories of individuals or family groups. Often two bats will fly relatively close together for much of the time, separating briefly to chase and catch an errant insect or two.

A countryman has always had an amicable relationship with these little leathery-winged mammals and has had some interesting experiences with them over the decades. One such experience entailed extricating one of the little creatures from a ribbon of flypaper hanging on the enclosed porch of the house. The porch door was always shut or the screen part only was open, but somehow the bat squeezed in overhead apparently.

With great effort and carefulness on the countryman's part and with many high pitched protests from the "flitter-mouse", it was finally extricated from its prison, cleaned up and released.

A lone peeper calls from nearby as it slowly works its way from the woodlot where it spent the summer to its winter quarters. Its shrill small voice pierces the darkness.

A man inhales the crisp night air slowly, enjoying its sweetness, as he watches the ever-enlarging silvery moon slip out of sight behind the western tree line.

# AUTUMN

The passing of summer is duly noted by the simultaneous arrival of the Autumnal Equinox. Six months ago, the twentieth of March, was the last time a countryman enjoyed days of equal daylight and darkness. Soon the golden days of October will be upon us, with ripened grapes and apples and many other fruits of man's labors ready to be harvested. There will be frost on the pumpkins and over the cornfields the leaves will be curled at the edges with brown contrasting with the yellow-green where Jack Frost has run his frigid fingers over them. But wait – September isn't over just yet.

Muddy footprints cross the macadam road from a nearby cornfield, clearly delineating the nocturnal activities of raiding raccoons.

Three ducks quack loudly in alarm as they rise from the farm pond at a countryman's approach and are soon lost to sight in the distance. Their pattern of flight mingles closely with that of five Canada geese approaching from the opposite direction.

For once the geese are flying quietly with no honks of heraldry, just the swish-swishing of their great wings as they fly low overhead.

A small family group of phoebes use the fence posts and wire adjacent the cornfield to launch forays on the many flying insects in and about the cornstalks.

Walking further afield, a man stops to watch a flock of cedar waxwings atop two adjoining trees as they also continuously fly out and back after insects.

This day is mild and the wind currents and weather must be just right for the migrating monarch butterflies. Scores of them can be seen within a short period making their way in a southerly direction. Some fly low, some are very high in the air, while yet others take a course in between.

Contrary to popular opinion, there is often much life in and about a cemetery. Beneath a hickory tree alongside the cemetery road are scattered the empty green hulls of hickory nuts. The squirrels have been harvesting them from their branches and husking them, allowing the hulls to drop to the ground below. If, after being husked, a particular nut is not good, the squirrel will let it also drop to the ground below, to lie rejected with the hulls. Occasionally, a good nut will slip from the paws of the squirrel and it too will fall down. A foraging chipmunk or another squirrel may notice it and after a careful inspection take it to its own place of storage.

A little red squirrel scampers up the trunk of a horse-chestnut and peeks cautiously around the tree bole at a countryman as he approaches then disappears among the heavy foliage.

Later, a deer and her two fawns make their way slowly across the roadway, pausing now and then to nose about the fallen acorns and the caps that have been separated by the wheels of numerous vehicles. Their tails switch nervously and perhaps to dispel biting flies as well. The fawns, now nearly as large as their mother, enter the brush lot first. The doe, by now, has noticed the countryman who is standing motionless, but is still not a normal part of the roadway. Seeing and scenting nothing to alarm her, she soon disappears into the undergrowth.

A chipmunk chucks from its seat atop a granite marker, assuring a man that there is, indeed, life even among the stony evidences of death.

October arrives on the heels of Autumn. The first frosts have already touched some of the rural valleys, a portent of things to come. There is a not-so-subtle change in the air – in the whole atmosphere. The wind shifts more to the north, bringing with it colder mountain air. A countryman can sense a definite shifting of the seasons.

Leaves are changing color more each day. Although the woodlots appear primarily green, there are splashes of bright colors to be seen and yellows accent the light greens of the willows and the walnut trees. More and more the sumacs are displaying their flaming pennants while the Virginia creeper vines wear wine-red robes decorated with blue berries as they clamber upwards into trees and over lower shrubs. Many birds are fond of the berries and help propagate the plants through their droppings. The vines provide not only food but shelter for nests of both birds and paper wasps which come to view once the leaves fall off the vines.

The once-brilliant fields of goldenrod are becoming subdued as the flowers dull to bronze and shades of brown upon approaching maturity.

While the goldenrod is in a state of decline, the New England asters seem to bloom more boldly each day. Great amethyst flower gems they are, their centers providing a lovely old gold contrast to the whole.

The geese are becoming increasingly restless and small to medium flocks build along the lakes and are seen and heard more often as they fly back and forth between them. Sometimes a countryman thinks that the geese are impatient for the great fields of corn to be harvested so that they can gather and glean what the harvesters leave behind.

On a mild autumn day the gentle breezes that waft over the vineyards bring the scent of sweet, succulent grapes. A man needs only his nose to tell him what time of year it is.

Jays fly across a sunlit clearing, some screaming apparent insults at others who are carrying acorns in their beaks to stow away in some secret hiding place. It will be a secret only if other sharp eyes do not observe where the acorns are placed.

Among the hickory and walnut trees can be heard the rasping of sharp squirrel teeth as they cut their way through the hard shells to the protein-rich meats within.

Summer's bounty has become Autumn's harvest. Man and creature alike are increasingly active as they gather in the products of the previous seasons.

A plump woodchuck waddles across the roadway to its burrow. In its own way it also has been "harvesting" and storing food away as fat all over its body. These fat supplies will sustain it during its long winter sleep.

There is an abundance of small fruits for the wildlife community to feast on as Autumn works its way into October.

The small, wild fox grapes hang in loose purple clusters. A man samples one of the darkest ones and finds that it isn't too sour, but very tart. The pioneers used them for jelly and some of our cultivated grapes had their origins from this little wildling.

Hawthorne trees are decorated with their red-orange "apples" and the ground beneath the black cherry trees is covered with over-ripe, wrinkled fruit. The elderberries are past their prime and are drying slowly on their stems.

Mullein spires are setting seeds with only a few yellow blossoms at the top of the stalk. When mature, the seeds and the larvae that dwell in many of the stalks will provide food for several species of birds during the winter months. Thistle seeds are ready to sail away beneath their parachutes, while brown burdock burs are almost ready to attach themselves to any passerby assuring themselves a free ride to destinations unknown.

Some days are sun-filled and mild while others are windy and cool or perhaps a cloudy day brings much-needed rain to refresh the dried earth. Some days, although warm and sunny, seem to be unusually quiet. A countryman has to go to the woods-edge or the hedgerows to hear the voices of nature and there are times when these places seem abandoned. Along a brushy roadside two robins chirrup as they join several waxwings feasting on some remnant black cherries clinging to their branches.

A female junco sits quietly in a bush and watches as several chickadees go trooping by. A nuthatch murmurs nasally to itself as it circles headfirst down a tree trunk. A flicker calls "clee-up!" from a precarious perch atop a sapling and a lingering catbird cries from the thicket.

Two members of the phoebe family still find insects near the corn field, delaying their southward journey. Six turkey vultures serenely sail overhead, their wings tilting and dipping as they search for and encounter the various thermals rising from the sun-heated ground below.

A black and white pileated woodpecker with jaunty crimson crest flies between woodlots with strong and steady wingbeats. It has been quite a while since a countryman has seen one of these crow-sized woodpeckers.

In wavering wedges and long lines, the geese are commencing their migration. Every day or so, small flocks are heard if not seen. Late afternoons as well as early mornings are good times to observe them.

The last long rays of the afternoon sun stream through the leaves of a Virginia creeper vine. It climbs like a living flame up a spruce tree to which its tendrils cling. A man watches the scene for several minutes until the rays weaken and the flame flickers and dies with the setting sun.

A lone cricket chirps slowly as a man heads towards home. There is a certain bite to the evening air despite warm tints of the salmon sky above the western ridgeline. A countryman has a feeling that, come morning, frost will be on the pumpkins.

Oh, the glorious golden days of October! Blue skies, warm temperatures, all are the essence of Autumn. Ash trees are dressed in mauve or sometimes yellow while yellow-green leaves of walnut and willow drift to the ground, twisting and turning in the afternoon breeze.

The maples blaze across the hillsides with their yellows, oranges and reds. The flaming foliage of the sumacs burns its way along the roadsides and hedgerows. In the aspen groves, leaves spill like coin of the realm on to the bosom of the earth, covering it with their translucent gold.

A vulture competes with a red-tailed hawk for the rising thermals in the blue vault above while closer to the earth dragonflies hover and skim about like miniature helicopters.

Purple pokeweed berries cling to their rose-colored stems. Native Americans, and later, the colonists, used the berries to make a vivid and quite permanent dye. Country women dreaded hanging their clothes on the line when robins and other birds were feasting on pokeberries.

Here and there in sheltered places butter-and-eggs greet a countryman as he passes by, their yellow and orange blossoms bright in the grass. A few wild morning glories still bloom beneath the grape rows amidst the scarlet leaves of the creeper vines.

Attached to a short green stem at the base of a dried thistle stalk, a brave little purple blossom is making the most of these warm autumn days until it either reaches maturity or succumbs to a black frost.

A flock of robins spend a few hours in the area and then are on their way again. A red-bellied woodpecker "churrs" in the woodlot as it inspects the limbs of oak, hickory, ash and maple. In the trees and sumacs along the hedgerow a diminutive cousin, a downy woodpecker, also seeks out food among the branches. On occasion, it gives its low whinnying call as it flies from tree to tree.

Bluebirds call sweetly from the pasturelands while the roadsides are host to song sparrows, flickers and a casual catbird that mews in a rather melancholy vein.

Traversing a trail through rather dense undergrowth, a countryman is accosted by a blackberry bramble with three large, fully ripe berries on its branches. These he consumes with no hesitation on his part. Then he notices that there are several more berries on the cane that are still red and will, no doubt, attain ripeness due to their sheltered location. This was certainly an unexpected but delightful highlight to his ramble.

Bats had lately been conspicuous by their absence in the evening sky but as dusk settles out across the vineyard a large bat makes its way erratically across the sky. Deer feed contentedly in a distant field and, as dusk deepens into darkness, a little house wren retires to the brushpile for the night. Cheerful crickets chirp and katydids rasp away against the stillness of the night.

Early autumn is the time for the conifers to shed their needles. To be more specific, the needles that turn rusty-red and shower to the earth are the ones that were newly formed in years previous. There is a continuing cycle of growth and discard as the new growth over-shadows the older and with the resulting lack of sufficient sunlight photosynthesis no longer functions, the needles turn brown and ultimately fall away to nourish the root system below.

The deciduous trees now are robed in their resplendent autumn garb. They are like a treasure chest overflowing with the leafy gems of topaz, ruby, garnet and emerald, mixed with garlands of gold. Among the oaks can be found the more subdued shades of old copper, bronze and dull reds.

Walnut trees stand stark, their limbs decorated with yellow-green ornaments, the result of their summer's labors. Gray squirrels can often be seen carrying the globular green-sheathed nuts to some secure place where they energetically remove the thick covering and then hide it away for safekeeping. Many young walnut trees get their start in life this way.

The canary-like voices of a goldfinch family come from a weed patch as they seek out ripened seeds. A song sparrow sings tentatively a few notes and springtime briefly returns. A robin carols for a few moments, adding to the sensation. Hidden in nearby bushes a white-throated sparrow sings its soft, sweet song and a countryman knows that it is winter and not summer that is on the way. However, the warbling of a bluebird quickly dispels any thoughts of winter.

Crows congregate about the cornfields and adjacent woodlots. Sometimes a flock of half a hundred or more will be seen. Perhaps it is a sort of family reunion at the end of the summer. Whatever the occasion, they have a noisy and sometimes rowdy time of it.

It seems to a countryman that many of the wild folk are taking time out for some relaxation, a holiday of sorts, if you will. The young ones have been nurtured and raised to maturity. Now individuals and family groups can enjoy a slower pace and, with some species, enjoy the companionship of one another.

The young turkeys have grown up and have learned to fend for themselves. A few of their number are missing, but that is the way life in the wild is. The alert and strong ones are those best fitted to maintain the species.

The mellow, yellow days of October are days to revel in They are days that are not too warm and are yet not too cool. They are robust days full of life even as signs of death are all about. The days of autumn are like our lives, brief perhaps, but meant to be enjoyed to the fullest.

Scattered snowflakes drifted earthward from the leaden, early morning sky. Along the eastern horizon a scarlet sun was masked by clouds, creating a saffron atmosphere that spread across the ridges.

Roofs were coated with a thin layer of wet snow; multi-hued trees stood, shivering in the sub-freezing temperature, their leaves coated with the white crystals. White-throated sparrows spoke with muted voices from the snow-dusted bushes. Birds gathered about the local feeders for their breakfast.

Late October is always full of surprises. Sometimes there is a brief burst of summer weather, or the fall rains get an early start. Then again, there may be the first hint of winter and the first snow.

Signs of an autumn in progression are everywhere. The gold of the goldenrod has matured into gray seed-fluff, while the amethyst asters no longer spread royal purple through the fields and meadowlands. Only a few scattered blossoms are now in evidence. Thistledown hangs from their prickly stems and stick-tites and burdocks are eager to hitch a ride from any passerby.

The frost-bitten countryside lies brown and sere. Deer walk through weedy fields, their new winter coats blending perfectly with their surroundings.

Bucks are in the rut and evidence of this is seen in the many buck "rubs." A countryman came across a couple that were interesting. The first one involved some small willows, their slender trunks were scraped free of bark and a branch nearly an inch in diameter was broken off. Tattered pieces of bark drooped forlornly from the mutilated trunks.

The elderberry bushes suffered a similar hormonal attack. Prior to this, however, the withered berry clumps had been harvested by a deer. The way they had been removed provided the necessary clue. The deer family has no upper incisors and they must tear their food off rather than cut cleanly.

Huge crow congregations gather in the woodland tree-tops. While some settle to rest, others circle the area, while still more arrive from all directions. Nearly a hundred may assemble at one time. It seems to be one of autumn's many rituals. After a very lengthy discourse, the meeting may adjourn and the whole group will take to the air. Once airborne, they will circle and swirl for many minutes before striking off for parts unknown.

Flying rather haphazardly over the vineyard, a great blue heron finally alights in the top of a walnut tree where it rests and surveys the countryside. A few minutes of rest and it took to the air once more, its great broad wings beating steadily, carrying it out of sight.

As a countryman observed and listened to all these signs of autumn, he could visualize October and Autumn walking hand in hand along the ridges and down the valleys, on their way to meet November.

The brightly colored leaves of the sumacs are gone for another year. The sumacs stand, darkly delineated against the sky, their new fruit clusters lifted like so many torches to brighten the way into and through the coming winter. Although often despised as a prolific weed tree that grows rapidly in often the wrong places, yet the humble sumac provides color and even food during the drab portions of the year. It provides sustenance for over-wintering birds and many songbirds find the fruit most welcome upon their first arrival in the spring.

In years gone by a man remembers going for enough of its dead limbs and trunks to make a quick, hot fire for a noon meal in the wood range. Country folk did this oft-times so as not to have a lasting fire in the kitchen during a hot summer day.

Warm autumn mornings, after a crisp night, are welcome to both man and wilding alike. Mourning doves, starlings and a robin or two sit, facing the sun as if to energize themselves for the day's activities. A man also is inclined to stand in a sheltered spot and absorb some of its warm, beneficent rays.

Squirrels scold and chipmunks chuck from the woods, the hedgerows and brushlots. The main rush of harvest is over and mild mellow days are an excuse for even the most industrious squirrel to take some time out to just sit on a limb and contemplate the world about him. Sitting with his front feet folded against his chest, and tail curled over his back, this usually active squirrel will soak up some sun. If he feels relatively secure in his position he may even catch a few winks, trusting his ears and other senses to alert him to danger.

Walking along a back headland, a man discovers a pile of black feathers. Something caught a crow off guard which, in itself, was somewhat of a feat. There was no blood, no sign of a struggle, just neatly plucked feathers. Something surely "ate crow" literally that day.

A little ways further and a man spied a bunch of grapes hanging beneath several shriveled leaves. Jack Frost's freezing fingers had barely touched them and they were still sweet and succulent. A man likes to see how far into the season he can find, in sheltered spots, fruits that have escaped the increasingly severe frosts.

An immature red-tailed hawk sails slowly overhead, its grayish tail fanned and guiding it along the upper air currents. On occasion an adult may be seen sharing the sky with it, no doubt one of its parents. However, it is now fully on its own and a man would hope that it will survive its first winter. This is always a crucial period of time for all young creatures and only the strongest and most alert ones will survive.

A female cardinal flies from a clump of willows to some sumacs where she sits demurely for a few minutes before continuing on her way. A man has not seen a cardinal for some time and was pleased to note this one.

November robins are always a joy to see and hear, even if their voices lack the spiritedness of Spring.

Juncos fly up from the road side, their white outer tail feathers flashing as they disappear into the bushes and white-throated sparrows whistle gently as a countryman walks slowly towards home.

Autumn rains, with their accompanying winds, have finally caused many of the leaves to relinquish their grip and take their places upon the earth's bosom, providing an insulating blanket from the cold to come.

The leathery leaves of the black oaks shroud their limbs with a cloak of rich rusty brown. Quivering in the wind, the leaves make slight, rustling sounds as a countryman pauses to look and listen. They may firmly maintain their hold well into the season or let go and fly away before some untimely wintry blast.

A red-bellied woodpecker makes its presence known with its characteristic call although it keeps from view for several minutes until it flies over a field to a neighboring patch of woods.

A man will ramble about at this time of year just looking and listening. Sometimes the only sound is the wind in the trees.

The pine branches make a soulful sound as the wind flows through, while the few remaining leaves left on the aspens make a noise not unlike the clapping of elfin hands. The spruces moan as the wind swirls through their stiff needles and a man remembers standing, many years ago, at the edge of a Maine spruce bog and listening to the same wind-song. He listens intently for several minutes now and the sound brings to mind memories that have lain hidden these long years.

A great flock of geese honk their way to and from the cornfields. Soon their supply of corn will lie several inches beneath new-fallen snow.

As a countryman walks slowly homeward he hears voices reminiscent of those New England years. They sound somewhat like those of English sparrows but of a timbre more vibrant and clear. The voices belong to evening grosbeaks and finally he spies one, sitting alone in a tall tree.

They are solid, chunky birds with large pale beaks, hence the name grosbeak. These seed-eating birds are dull yellow and tan with black wings and white wing patches, while the females are silvery-gray with enough of the yellow, black and white to be recognizable. They drift down from the north country in late autumn and early winter. They become very numerous during those years when the cone crop is sparse or non-existent in their northern habitat.

A countryman reaches the shelter of his porch just as a passing shower quickly dampens the ground. A short time later the sun re-appears, low in the western sky. A man steps out into the open and, just as was anticipated, a rainbow arches half-way across the dark eastern sky. Its vibrant colors slowly dim as the sun slowly sinks from sight, leaving behind contrasting bands of pale yellow clouds and aquamarine sky.

A man continues to watch as the color fades from the clouds. As he is about to go inside he notices five ducks hurrying homeward as the shadows lengthen, their forms etched darkly against the now deep blue sky.

Hickory and walnut trees stand, leafless arms outstretched, naked in November sun. In contrast, the chokecherry trees remain robed in colorful shades of rose pink and reddish purple that seem almost luminescent in the bright morning light.

Narrow, yellowish leaves drift down from the big willows and a nuthatch explores one of their rough, convoluted limbs talking to itself in sharp nasal notes.

Cirrus clouds, thin, wispy and feathery, flow across the sky like great brush strokes on an azure canvas.

The canary-like voices of goldfinches are heard along the field edge where they are feeding. Garbed in their winter suits, they are difficult to locate amongst the weeds until one flutters from one weed stalk to another. They converse contentedly as they forage in the autumn sun.

A few robins sit scattered in the tops of saplings, their brick-tinted breasts being warmed by the sun. Several bluebirds wander about the area, their voices bearing a hint of mournfulness. Perhaps it is because of the heavy frost which has so recently melted and is dripping from the tips of branches.

Mourning doves fly up from the road-edge, their wings whistling as they arrow away. Large flocks of geese move from cornfield to cornfield, their clarion voices adding to the sense of a season drawing to a close.

A countryman scuffs noisily through a wide patch of leaves beneath an old sugar maple, not caring who hears his approach. He enjoys the crisp crackle of the dried leaves and, most of all, the spicy redolence that is created by their disturbance. He knows that all too soon they will lie sodden and silent, their now colorful forms blending slowly into the earth and contributing nutrients that will help to nourish the next generation.

As a man ambles along he finds himself confronted by a bull thistle of giant proportions. Its gaunt, dried and prickly stalk reaches upwards a good eight feet. Numerous seed heads droop limply from their branches. Its roots most certainly had reached an abundance of nutrients and moisture in the previous months in order to attain such stature.

A small spruce tree has captured colorful leaves of maple, ash and chokecherry and holds them carefully in its bristly boughs. The leaves contrast brightly against the dark green spruce needles.

Milkweed pods burst down their middles and silvery shrouded seeds escape to be caught away by any vagrant breeze. On any windy day their numbers may be seen floating away. Some get caught almost immediately in nearby tall weeds while others float across fallow fields. Others rise to great heights and catch a ride on strong air currents that will carry them to distant places.

Nice days in November usually are few in number and a man has early learned to appreciate each one. All too soon gray, colorless days will arrive, their clouds bringing the autumnal rains and scattered showers of snow. This is as it should be. This is a vital part of the circle of the seasons and the earth waits patiently.

The long lake, hidden below the ridges, makes its presence known on a frosty morn by great banks of clouds hanging over it. The waters, warmer than the surrounding air, create their own weather. In this instance it is moisture which earlier in the morning may well have been called fog but in the early morning sunlight it now appears as an elongated bank of clouds stretching north and south above the lake.

Nearly a foot of snow blankets the ground, the result of an early storm. Deer trails criss-cross the landscape while crow tracks wander across vineyard headlands, connecting the adjacent sections.

Along the edge of the meadow, juncos and sparrows have left their diminutive tracks in the snow, stitching together the innumerable patches of weeds upon which they have recently been feeding.

A Cooper's hawk files with intermittent wingbeats from woodlot to woodlot and disappears among tangled branches in search of unwary or careless birdlife or small mammal.

As the day wears on, the air becomes mild and one of a small number of robins chirrups several times. As a countryman listens, he can imagine for a few moments that it is March instead of a few days short of Thanksgiving.

However, the canary-like voices of the goldfinches in the nearby hedgerow are not of the same tonal quality as they would be in that late-winter month. Their voices tend to denote impending rather than declining winter. So it is that the momentary magic spell is broken. Even the call of the flicker lacks the warmth and the energy that would be present in early spring.

Like it or not, winter lies just ahead and with all the snow now covering the landscape one might easily assume that it had already arrived and wasn't soon to depart. The weather this time of year is quite fickle and bare ground may soon appear once more. And then again, it might not.

A mourning dove sits somberly on a power line, its head hunched into its shoulders and its feathers fluffed against the chill air, the result of a slowly disappearing sun to the west.

As a countryman walks along a country road, a nearby cornfield stands in close ranks like a battalion of khaki-clad soldiers at attention, their forms stiff and erect. A light wind moves through their ranks and corn leaves rustle and rattle with its passing.

Approaching a thin stand of pines, a man hears a shrill whistle as a red-tail launches out from its perch in one of the conifers. Its piercing cry is heard once more as it approaches a distant woodlot.

The sun sets, the orange globe sinking behind blue and purple ridges that soon are outlined darkly against the salmon sky.

Walking past the farm pond on his way home, a man is startled by the piteous cry of a killdeer. It rises on slender, pointed wings from the muddy patch of ground just below the pond, circles briefly then returns to settle once more on the wet ground, uttering again its soulful cry.

A countryman cannot help but wonder what has kept this bird from going south before this and only wishes the best for the lone little plover.

Gray November days become more frequent as it draws near the close of the month. On occasions the sun will peek faintly out from behind the clouds or through thinning overcast. Temperatures hover between the mid- to upper twenties and low to mid-thirties.

On days that get above freezing the snow settles and melts enough to keep the streamlets flowing. They gurgle and splash as they follow paths of least resistance, making merry melody in their passing. A countryman will pause for a few moments to listen to their soothing song.

A pair of cardinals flies low across a small clearing to a brushy tangle overhung with the vines of wild fox grapes. The small blue clusters dangle from the vine enticing the redbirds to sample a few. Several bluebirds already have been tasting the tart treats and warble softly.

Two white-throated sparrows sit quietly in the middle of a bush watching the activities of their neighbors. A noisy group of evening grosbeaks perch in a tree-top a short distance away.

A ragged vee of geese honk their way across the leaden sky, no doubt intent on finding a cornfield which has not already been gleaned. A man thinks that they should go south instead of depending on their fortunes.

Now that the trees have lost their colorful cloaks various summer homes have been revealed. Stick nests of jays and crows are common. Then there are the carefully woven nests of goldfinches in sapling crotches, while lower down in the thickets are the delicately-designed homes of the yellow warblers and even lower those of the black-masked yellowthroats. Their voices are no longer heard but evidences of their activities remain.

In places least suspected during the warmer months hang the gray paper nests of white-faced hornets. They hang empty now, swaying in the brisk western breeze. Some have already been ravaged by eager bird beaks intent on finding a hidden morsel or so.

Near the tops of many trees are the large, leafy summer squirrel homes. These were airy, well-ventilated, yet snug structures designed to provide secure yet cool comfort during the long hot summer days. Tree cavities lined with aromatic dried leaves and grasses now provide the bushy-tails with winter quarters secure from cold autumn rains as well as the icy blasts and driving snowstorms to come.

During this day's ramble a man sees numerous tracks in the softening snow. A skunk ambles along nosing now and then into a grass clump hoping for a mouse. The star-shaped prints of an opossum wander aimlessly about the country-side. Crow tracks mingle with those of meadow voles in the vineyard, its owner in hopes of a mouse meal. Deer were on the move during the night and pre-dawn hours and white-footed mice have scampered all about the weedpatch at the edge of the brushlot.

Overhead, a pair of red-tails cruise on an air current. A crow sees them and cries "hawk! hawk!" as they pass its perch but makes no move to harass them.

By late afternoon it has become very foggy. Even trees close by are only faintly seen. Robins chirrup from somewhere in its nebulous depths, one bird even briefly caroling as if to dispel the gloom. Jays call in harsh tones while in the far distance a crow caws in somber tones. The fog thickens and swirls. A chickadee and a nuthatch exchange viewpoints as they forage in the thick grayness that mantles the land.

A countryman heads for home, leaving the land and its inhabitants to the gathering dusk of evening and to the fog.

Gray November days merge in an unassuming manner into those of December. There are spells of sunny weather interspersed by frequent rains, cold snaps and snow squalls.

During the milder periods cedar waxwings may be seen and heard as they feed on frost-bitten small fruits or launch out into the air after insect prey that has so far survived.

Two male downy woodpeckers inspect the branches of a clump of sumacs, keeping a short but respectful distance between themselves. Juncos busy themselves in the weed patches close by the woodlot. A robin "tut-tut's" sharply while eating fox grapes. A male red-bellied woodpecker clambers about the thick limbs of a sturdy old oak tree. It calls vigorously several times as it continues its explorations. Two male flickers forage together, first in the big willows, then over in the sumacs to sample the tart fruit. Then they are off, in undulating flight, to the woods where their petulant calls can be heard.

Male birds no longer are competing for territories and this is why they may frequently be seen together during the colder months.

For several days a male downy woodpecker spent his spare time excavating a new home for the winter. His choice was a dead but firm limb jutting out from the trunk of one of the large black willows. He started the hole from the underside of the slanting limb, affording maximum protection from wind and weather. The last a countryman saw of him his project was nearing completion. He would disappear into the ever-enlarging cavity, his labors easily heard, although muffled. Soon he would back out with a beak full of chips. This was repeated two or three times, then he would resume his chiseling.

During the frequent somber days, a man must be content to listen for rather than look for the wild folk. Sometimes many minutes will go by before any sounds of life may be heard. Hidden by thick bushes, underbrush and tree limbs, the wildings go about their business unobserved.

Then a note here and a voice there may be heard. Evening grosbeaks may be the first to break the silence or it may be a robin hidden in the sumacs. A jay imitates, rather lamely, a red-tailed hawk, but a man knows that it is only a jay playing jokes on its neighbors. High up among the branches come the nasal notes of a nuthatch and farther away the whinnying call of a downy woodpecker. Bluebirds can be heard but not seen as they fly high overhead. A chickadee calls cheerily nearby. These are some of the sounds that may be heard on a late autumn day.

As is his custom, a man pauses on his way home to listen to the liquid melody sung by the little stream. He listens intently as it sings a song of stones, of quiet pools and miniature waterfalls. The water gurgles and tinkles and splashes and a countryman listens to a song born of snowmelt and recent rains.

Fog thinly veils the land as a countryman starts out from home. Nothing stirs; it is as if all life forms have vanished. He walks out the vineyard road, the honeycombed frozen earth crunching crisply beneath his feet. A narrow band of snow thinly covers the ground beneath the grape vines. Then, a crow calls. Another responds in the distance and the silence is broken. Soon, several are calling back and forth and it is apparent that something has attracted their attention.

After a short time they disperse, each going his separate way, and a man is once more left alone with his thoughts.

Approaching a watercourse, a man notices deer tracks in the now-frozen wet earth. There are the heart-shaped tracks of a doe and a yearling and, farther on, the broader, sometimes splayed hoof prints of a buck show in two or three places.

A fallen tree trunk at the edge of a hedgerow reveals the passage of several small animals over its snow-dusted surface. Only the tracks of a gray squirrel can be clearly identified as the tracks are superimposed upon each other.

A countryman reaches a road and continues along it. As he walks along he hears a rustling, crackling sound off to one side. He stops and listens a few moments. The rustling continues and then a movement in a medium-sized spruce reveals a gray squirrel with an ear of corn, still enclosed within its husk. The squirrel seems undecided as to what to do with it. Finally, grasping the ear firmly in his mouth, he descends the trunk, the protruding ear bumping into branches all the way down. Soon the squirrel can be heard bounding through the leaves of the nearby woodlot with his booty.

A bluejay screams from its secluded place in a thicket, then all is quite once more.

The little brook sings its song as it leaps over ice-encrusted sticks and small branches that lay across its path. Thin layers of ice crystals reach out from each side of the stream until halted by the current.

Slowly, a countryman retraces his steps homeward. Silence hangs about him as he leaves the land to the crows, the squirrel, the jay and to the fog.

By mid-December, Autumn is rapidly drawing to a close and a new season is about to begin. Some years it would seem that winter had already officially begun. Snow storms have left deep, white blankets upon the land. There have been snow squalls that have swept over the ridges, across the vineyards and the lakes and down the long valleys. There may even have been brief periods of freezing mist or drizzle that would sheath branches and weed stalks with shining icy armor.

Then there are the rare, bright and beautiful days that remind a countryman of early spring. The birth of one such day was recently witnessed by a countryman during his morning walk.

The morning was cold. The mercury dipped overnight below the freezing mark, not much, but combined with a light wind, enough to be noticeable. A thin layer of clouds lay across the eastern ridges while overhead the sky was relatively clear. High in the western sky hung a silvery moon. Only two days past full, yet already one side was diminished enough to give the silvery circle a somewhat lopsided appearance. Below the moon and hovering over the western ridges were more thin clouds.

By now the eastern sky has become a pastel yellow-orange, heralding the imminent dawn. Many minutes go by and the sky deepens to orange in one area. A broad streamer of light flares upward here.

A crow caws and soon two fly over, silhouetted against the aquamarine sky. A jay screams nearby, a nuthatch calls, and in the distance, a red-bellied woodpecker and a flicker each voice their opinion of the dawning day.

The eastern sky becomes increasingly brighter as the earth turns. The western clouds have become tinted with a shade of rose tinged with lavender but start to fade even as the eastern sky brightens. A man turns his attention now fully upon the eastern horizon and watches until a brilliant spot appears through the distant trees and then, the flaming disc seems to rise swiftly into the morning sky even as a man quickly averts his gaze.

Within minutes, the sun is well above the ridge-line and almost immediately the wind picks up just as a new day is born.

A goldfinch dips and rises in flight overhead, and utters a few canary-like notes in passing. Several house finches sit in a bush chatting amongst themselves. They fly up as a countryman approaches as do also several juncos a few moments later as he walks down the narrow trail past the brook.

Two bluejays busily sample the tart fruit as they sit in a sumac. In the distance, a robin "tut-tuts" from his perch atop a spindly sapling. His brick-red breast shows well in the morning sun. A bluebird warbles softly in flight overhead. The temperature is steadily climbing in concert with the sun.

For a while it is difficult to realize that the Winter Solstice is less than one week away.

# WINTER

The Wheel of The Seasons has now turned through the Winter Solstice and already the days are imperceptibly growing longer once again. That doesn't mean but what there are plenty of short days and long winter nights ahead. What it does mean is that the balance scales of the season are tipping more each day now in favor of Spring.

For the next three months there will be many interesting things happening in the outdoor world that occur only during the winter season. For now, a countryman will take one day at a time, whether it be sunshine, sleet, freezing rain or wind-blown and drifting snow.

A bright red male cardinal is a splash of color against the grayish background of the branches of the bush in which he quietly sits, surveying the world around him.

Several chickadees do acrobatics in a nearby tree as they search for food. House finches huddle in a large shrub, chirping back and forth in soft, musical tones.

A gray squirrel balances precariously on a slender branch before making a long leap to another equally slender branch of a neighboring tree. The branch droops suddenly beneath the squirrel's weight, but just as quickly the squirrel climbs upward to more secure footing and is soon seated where the branch joins the trunk of the tree.

For several minutes the squirrel sits quietly in seeming contemplation. A hind foot reaches out to scratch at an irritating flea. The other hind foot repeats the process. Then a face seems to be suddenly soiled and that is dutifully washed. The tail is next, carefully washed, combed and fluffed all in several very dexterious movements. A man watches all of this in bemused silence.

The wheezy, chickadee-like notes of a tufted titmouse betray its presence. It flits here and there, inspecting twigs and branches for insects, their larvae and anything else that might qualify as food. Working along a branch, its bright shoe-button eyes notice a sunflower seed wedged into a fissure of the bark. How it had gotten there is anyone's guess. Perhaps sometime in the not too distant past the titmouse had hidden it there and now decided it was time to retrieve it.

Holding the seed firmly with its strong little toes, the perky titmouse worked the nutritious seed, piece by piece, out of its shell and devoured it. That accomplished, it flew off into the brushlot.

A man heads for home beneath a darkening sky. From all appearances, it would seem that a storm of some sort was in the offing.

Our calendars tell us that we are already into the first full week of the New Year. Nature's calendar has a different perspective of things however. The last week or so of December may or may not be any different than the first few weeks of January. The wild folk pay attention to their calendar of the seasons and conduct their lives accordingly. They react to mild or stormy days during the first month the same way they did during the last month. They will continue in this way until the days become longer and milder and their biological clocks tell them that it is time to think of territories, mates and nests once more. The mockingbird seen on Christmas Day may well be seen again during a mild period later in the season. So too may the waxwings, the robins and even a bluebird or two briefly make an appearance, depending on the severity of the season.

The winter woods seem empty at times. Only an occasional woodpecker, nuthatch or a few chickadees or a jay or two can be seen or heard. Of course there are the ever-present crows, but there have been a few times when they too have been conspicuous by their absence.

Sudden cold snaps drive the wildlings into dense cover out of the biting cold, the roaring wind and driving storm. Even when the storms subside, the cold may well remain or even intensify. It is then that the countryside may appear devoid of inhabitant.

A few tracks tell of rabbits and their activities, as do those of crows whose three-toed prints stitch the vineyard headlands together. Serpentine ridges along the snow surface show where shrews have tunneled about in their eternal quest for food.

A young red-tailed hawk perches in a tree, only its head moving in keen observation of the ground beneath and the area about it.

A Cooper's hawk flies low across a fallow field and alights half-way up in an oak whose still-clinging leaves partly conceal the raptor.

In a woodlot, adjacent to a grown-over orchard, a little red squirrel scampers across the snow, up the trunk of a white pine and disappears into the dark green branches.

A flicker calls rather absent-mindedly as it hitches up a dead stub protruding from the main trunk of an old and gnarly-branched maple. Chickadees converse in low tones as they work their way through the bushes.

All of this may be seen on a mild winter day, with the sun shining brightly, albeit weakly, from its low position in the winter sky.

Another day, when the thermometer couldn't manage to quite reach into the 'teens, there was almost no activity that was readily observable. Out for a short stroll, a countryman found the countryside quiet except for the light wind in conversation with the tree branches and the powerline wires. Three crows were perched in a tree, sitting on their feet with their feathers fluffed for added insulation against the biting breeze. They sat, silent sentinels of the season, seemingly with nothing to say.

A flock of starlings appeared out of nowhere and as suddenly disappeared into the distance, leaving the land as empty as before. Only the crows remained, huddled on their branches and having nothing to say.

January thaws are unpredictable. They can come anytime during the month and sometimes they don't arrive at all. A thaw in the early part of the month is quite unexpected since by this time everyone has become used to winter weather.

With a thaw in progress, a countryman will be especially watchful for birds and other creatures that are normally not observed during the month and he is seldom disappointed.

Two female bluebirds sitting in a sapling, softly conversing with one another and a scattering of robins over the countryside are two examples. Then there was the mockingbird seen foraging for food or just sitting silently atop a large shrub. A wooly-bear caterpillar had gotten out of bed and was taking a leisurely stroll when the temperature rose above sixty one bright, sunny day.

House finches and goldfinches roamed the area in small, congenial flocks, their musical notes adding to the already pleasant day.

A dull red female cardinal flew past a sumac where a male flicker was tasting the tart fruit and found it much to his liking.

Over the course of several days of thaw, the normal winter birds will be seen or heard, such as downy woodpeckers, nuthatches, chickadees and a raucous jay or two. And of course there are the starlings, house sparrows, and always the crows.

As if there wasn't enough of "spring" in the air, several flocks of resident geese move, morning and evening, back and forth between the lakes. Their honkings even have that aura, that excitement, that they exhibit in the springtime. It is as if they could bypass winter and induce Spring just by their clarion calls.

Mourning doves sit somberly on powerlines or fly about the cornfields in their quest for food.

The tree sparrows finally have arrived from their northern homes, their tinkling voices now added to that of chickadee and nuthatch and junco.

Under the onslaught of southerly winds and occasional sunshine, the deep coverlet of snow left by the last storm quickly disappears. Only on northern exposures and other sheltered or drifted areas does any amount of the frozen flakes remain and even there the snow shrinks with each passing hour.

Snowmelt drips, trickles and then runs as it leaves its resting places to merge into larger and still larger streams as it hurries towards its final destination.

Later, cold rains add to the diminishing effect as the thaw draws to a close and the temperature drops and then hesitates just above the freezing line.

This brief, spring-like interlude has been appreciated but now it is time for frozen flakes to fall, covering the bosom of the earth once more in an ermine blanket and lulling the land into peaceful repose.

A countryman remembers years long past when the ground would be frozen by mid-November. The first snowfall or two may not have lasted long, but after that the snow that came remained until the following spring.

Over the years weather patterns have changed dramatically in some areas and one isn't quite sure what to expect other than that there will be periods of cold and snow for at least a few months.

The ground is finally well frozen. After the thaw and the subsequent rains, the mercury dipped well below freezing and the wind blew in frigid blasts, crystallizing the moisture trapped in the soil.

Perhaps, a man thinks, we can now get on with winter in a proper fashion. Only time will tell what vagaries of weather will be forthcoming.

Walking briskly over the frozen ground, a man looks and listens as scattered flakes of snow drift silently and slowly earthward. His course of travel takes him by the place where catnip grew rank and tall during the previous summer. He pauses and looks closely at the spots where creamy blossoms atop sturdy stalks were once alive with honeybees. Now there is little left but a few weathered and broken stems, beaten down by snow and rain and decay. Upon closer examination, a man can see small clumps of green nearly hidden beneath mats of fallen leaves. He plucks one of the small, tender leaves and crushes it between thumb and forefinger, releasing the aromatic scent that is so attractive to most members of the feline family. A man savors briefly what, to him, is a remembrance of days of warmth, of bird-songs and the hum of bees.

A man's thoughts are brought back to the present as a red-bellied woodpecker calls from close by. Scanning the trees with his binoculars, he soon sees a female woodpecker inspecting rough-barked maple limbs for appetizing items of food.

Continuing on his way, a countryman reaches the small grove of spruce trees. They have grown considerably during the past six years and have provided food and cover for numerous species of songbirds, some of which have nested in the dense branches. Deer have occasionally bedded down beneath their sheltering boughs.

The snowflakes thicken and the bare, brown ground slowly takes on a lighter hue. As a countryman stands beneath some of the sheltering spruce boughs he watches the fragile flakes collect on nearby bough tips. One of the boughs is at eye level and provides an opportunity to closely examine the delicate and intricately structured crystals.

As he watches, five flakes land together in a line on a bough tip. Five six-sided crystals that are so similar yet unique in itself and diverse from each other. Always six-sided, never five or seven or even in the form of an octahedron, but always hexagonal in shape and never quite the same in structure.

Observing the intricacy of structure and diversity of snowflakes will lead any sensible and thinking person to realize that this could never occur by mere chance. Snowflakes, like so many other things in the realm of nature, can be explained by men of science but can only be constructed by the hand of the Creator.

As the first month draws to a close a countryman suddenly realizes that one third of Winter has already passed. As he contemplates the passage of time, he slowly makes his way along a woods-edge and across a portion of field towards another woodlot.

A red-tailed hawk leaves its tree-top lookout as a man approaches but alights again after only a short flight of fifty yards or so, apparently not unduly alarmed at his presence. Carefully folding its broad wings and waggling its rusty-red tail, it continued its watch over the countryside, its backside towards the human who apparently meant it no harm since he was walking at an angle away from the hawk.

A few saplings and scrubby sumacs clumped together a short distance out in the field. Three bluebirds, two males and a female, perched here and there in the trees while nearby a crimson cardinal perched in silence as the bluebirds foraged in the area.

Continuing on his way, a man noticed meadow mouse runways in the snow connecting short tunnels with their diminutive footprints.

Rabbit tracks zig-zagged from the woods out into the meadowland. Registered in the snow were short hops interspersed with longer leaps which were made from excitement, play or perhaps momentary fright over some danger, real or imagined.

In his rambles further afield a countryman comes across a fox trail as it purposefully leads across yet another field, connecting two adjacent woodlots.

Along a gulley edge, squirrels have been active, apparently engaged in uncovering and retrieving previously hidden caches of food. Every so many feet a hole appears in the snow revealing the dried brown leaves beneath which also have been disturbed down into a shallow depression in the ground where the cache had been.

A pileated woodpecker flies overhead on great black and white wings and disappears into the woods on the other side of the gully.

Musical twitterings attract a man's attention to a small flock of birds in a tree a short distance away. The binoculars reveal several female house finches and a couple of males with their dull red breasts. Interspersed among the finches are several goldfinches in their winter suits. During the winter months it is not uncommon to see two or more varieties of finches traveling together in small flocks since their food requirements are the same or very similar.

Closer to home once more, a countryman sees the mockingbird perched atop one of its favored bushes. It sits quietly down over its feet and gray feathers fluffed against the cold.

Perhaps, a man thinks, there may be a mockingbird family in residence come spring. At least it is something to hope for as the weeks of winter continue to pass.

A winter that has more mild days than normal is certain to have some influence upon the activities of the wild folk and, to some extent, upon their environment. Wet areas very close to streams show an unseasonable shade of green for one example.

Of late, the mockingbird has been seen almost on a daily basis and two or more pairs of bluebirds are frequently seen and even heard on occasion. Unaccountably, robins have recently been absent from the area.

One day a raccoon was seen curled up in a large tree cavity, its backside exposed to the warming rays of the afternoon sun. It was indisposed to be disturbed even by the various vocal efforts of a countryman. In retrospect, a man considered that, if he was in that same situation, he would refuse to be disturbed also.

Two small, polka-dotted feathers clung to a low shrub. They moved slightly in a momentary whisper of a breeze. Upon further examination, it was noted that they were tinged with very pale yellow and the tiny central quill was nearly a lemon yellow. These observations confirmed a man's suspicions that these were breast feathers from a flicker which no doubt had fallen victim to some predator such as a sharp-shinned or Cooper's hawk since these species have been seen actively hunting in the area most of the winter.

Tree sparrows gather in the weed patches and busily feed upon the seeds of last summer's blossoms. A scattering of juncos venture among them while a man watches, flashing their white outer tail feathers as they move about the weed stalks or fly up into a nearby sheltering shrub.

In the top of a tall tree two bluejays and a male cardinal sit silent and serene, their colors contrasting with the winter landscape.

Some distance away, two more male cardinals seem to have conflicting interests as they chase one another through shrubbery and across the intervening clearings.

Over the wide expanse of a large field a rough-legged hawk is seen alternately flying and hovering on broad pinions. It is similar in size to a red-tail but its flight pattern and distinctive coloration is vastly different.

White "windows" show near the tip of each wing while its white tail is terminated by a broad black band. In searching for prey, it hovers on steadily beating wings much like an over-sized kestrel or sparrow hawk.

The tundra of the far north is home to the rough-legged hawks, but come winter they drift down into the northern and central portions of the United States. Seldom does a winter pass that a countryman does not see one or more of these beautiful big buteos hovering over a wide field in search of voles and other rodents.

The appearance of the rough-legs and other far-northern species are eagerly anticipated by a countryman each year. After all, there is much to be said in favor of Winter.

The sky is overcast on this February morning and the sun glows weakly, its edges fuzzy with diminished warmth. The mockingbird sits atop one of his favored bushes, feathers fluffed for added insulation against the morning's cold. A female bluebird perches quietly in a poplar tree nearby and overhead a small flock of geese head for the lake, carrying with them an apparent congenial conversation. The previous evening they, or some of their kindred, flew over well after dark and boisterously discussed the day's activities.

The countryside is relatively quiet and even the crows seem to have little to say. A gray squirrel scampers across a small clearing, its tail flowing behind it like a silvery-edged banner. With energy it climbs up a tree in the hedgerow and makes its way through the adjoining branches. Its movements are almost fluid as it flows from one branch to another and a man thinks that a squirrel in its arboreal environment is like poetry in motion. Despite certain short-comings of the species, a man can't help but admire and, yes, even like squirrels. He has had them for pets and has tamed some of the wild ones. He has had wild red squirrels eat from his hand while he stroked their backs and sides gently and all the while they would continue to eat unperturbed.

Another day the temperature is in the upper thirties and the sun shines brightly through scattered cumulus clouds that float across the azure sky. Opossum tracks wander over the snow, some of them a bit muddy after the animal had crossed a recently thawed area.

The male cardinals are becoming increasingly vocal as the month advances. One perches high in a tree where the slight breeze at its back lifts its crimson crest to an even more jaunty angle than usual while its clear, sweet voice rings out over the countryside.

Two nuthatches squabble briefly when their search for food brings them both to the same position on a rough-barked limb. With slightly raised wings and fanned tails they momentarily square off until one of them flies to an adjacent limb thus ending the argument. Soon one calls "ha-ha-ha-ha" as though there had been no disagreement at all.

The "knee-deep" calls of a chickadee join the laughter of the nuthatch and together with the song of the cardinal there emerges a medley that has the essence of early spring rather than that of early February.

As the days continue to lengthen, slight stirrings in all of Nature can be noted. Some are obvious such as the bird-songs and slightly swelling buds and certain animal activities. Others are less obvious and even unobservable, taking place beneath the ground or out of sight elsewhere.

Yet, before the real awakening can commence there remains another month or so of repose with only now and then a brief respite from the cold, the wind and the snow.

A week of unseasonably warm weather is enough to stimulate the local bird population into thinking there will be an early spring. That remains to be seen. In the meantime, woodpeckers drum for hours at a time, hoping to inspire a mate. If she is wise she will bide her time a few more weeks or so. Male cardinals compete for the best voice award and any judge would be hard put to declare a clear winner.

On a sunny morning, from three distinct locations, come the cheery calls of tufted titmice. This particular morning they all are calling in a different vein than earlier in the season. They are de-emphasizing the "r" at the end that is so apparent when they go about looking for "Peter" and instead are saying "petah, petah, petah". Then one of them starts to use the rather wheezy chickadee-like call that is part of their repertoire. This is only natural since they belong to the same family as the chickadees with similar habits and tastes. The titmouse however is a more retiring bird than the vivacious chickadee who seldom seems to allow anything to repress its natural buoyancy.

A robin "tut-tuts" in a subdued way as it enjoys a short repast of small, dark berries. This is the first robin a countryman has seen in several weeks.

A red-bellied woodpecker stops briefly to inspect several limbs on a large willow and then continues on to some more distant destination.

Crows gather in small flocks in adjacent woodlots, calling spiritedly back and forth about something that is important only to crows. Now and then several will leave one of the flocks to join one of the others. Their smaller cousins, the bluejays, have increased their activities the past few days and have become more vocal as well.

Two beautiful cock pheasants forage along a roadside but run quickly for cover as a man passes by. A recently arrived red-winged blackbird is perched in the top of a tall tree and gurgles “o-ka-lee” several times before flying away. A man considers what he has just seen and heard and decides that one red-wing doesn’t make Spring. Nevertheless, with what he has been hearing the past week things seem to be starting to add up in that direction.

The voices of springtime have been heard and yet February is only half over. A man will not complain but is content to listen to the varied voices that have so long been silent. Even as he contemplates the events of the day, several bluebirds warble softly as the shadows lengthen and they seek an evening meal to sustain them during the cold night ahead,

Two robins chirrup a pleasing duet before they too retire for the night, their voices adding to the illusion of spring.

The sun sinks westward, tinting the clouds a rosy hue. Even as the sun bids farewell to the day a yellow moon emerges from a cloud bank just above the eastern horizon. A countryman watches as the moon, in its fullness, slowly rises into full view.

Between the dark line of the horizon and the lower edge of the lunar disc a red-tailed hawk flies towards its place of retirement for the night. A man also turns his steps homeward after taking one last look at the beautiful scene that has so recently transpired before him.

After a day and night's rain the ground is saturated with water and the brooks are running full. Globules of water hang from every branchlet; weed stalks are dark and heavy with moisture. Fog clouds hang briefly over the lake then rise to merge with wrung-out cloud layers scudding across the sky.

A giant mullein, once green and standing straight and proud, leans to one side its great velvety leaves now shriveled and sodden by the recent rains.

Beneath the leaden skies a hunting hawk silently searches the earth below for something to ease its ever-present hunger. As winter drags on, its food supply steadily diminishes and will continue to do so until Spring arrives and with it an abundance of new life.

A skunk left its tracks in a small, remaining patch of snow. Could it be a coincidence that they become amorous about the same time that humans celebrate St. Valentine's Day? Probably so, and several other species are now being prodded into increased activity by similar inner urges.

Two chickadees call "knee-deep" across a small clearing and several robins chirrup inquiringly. One robin actually carols for several minutes and the illusion of spring returns despite the season. The illusion is further enhanced when a song sparrow bursts briefly into song from the top of a rain-drenched bush. This is the first song sparrow to put in an appearance since last autumn and a countryman wonders what other unusual things will occur this year, the year that is touted by weathermen as being so drastically affected by

El Nino. Time alone will tell. Meanwhile two bluebirds call sweetly as they fly overhead.

During the course of one of his rambles, a man spies a polyphemus moth cocoon only slightly attached to a small shrub. He removes it carefully and is immediately surprised at its lightness. Tipping it first one way then another he notices that there is no movement of material of any sort inside. Having opened many such cocoons on previous occasions, a man surmises that the larva had fallen victim to a braconid wasp which had lain its eggs either upon or within the caterpillar shortly after its last moulting of skin. Apparently the caterpillar was able to spin its cocoon just before it was consumed by the tiny wasp larvae. This being the case, there would be nothing left but a finely constructed but empty cocoon. The wasps would later exit through inconspicuous holes they would chew through the walls of the cocoon.

No beautiful big brown and tan moth, touched here and there with ochre hues, would emerge to enrich a warm June evening.

Lost in thought, tinged with a hint of sadness concerning the empty cocoon, a countryman turned his steps homeward. He stopped and "glassed" a distant field. Five deer and thirty-one wild turkeys were searching the field for sustenance. Here was seen the fact that even where there is death there also is abundant life in the cycle of the seasons.

Just the day before the storm several flocks of migrating geese flew northward, high overhead. Their honkings drifted earthward, muted by the distance. Often geese will move ahead of a storm before continuing on their way.

The mockingbird sat quietly in one of his favored bushes as has been his habit of late. Bluebirds warbled and a robin caroled cheerily. A dove sat sedately on a powerline wire, quietly observing the world around it.

Two cardinals called back and forth while in the distance a gathering of crows appeared to have a heated discussion of some topic best known to themselves.

Clouds increased steadily during the afternoon, obscuring the sun until the temperature slowly dropped out of the mid-forties.

Late that night wet flakes of snow started falling, slowly at first and then picking up in intensity. All night the snow continued to fall and by morning several inches of heavy, wet snow blanketed the earth with ermine. Throughout the day the snow fell, at times slacking off and then increasing in tempo.

Meanwhile the wind had increased with heavier gusts that dried out the surface snow so that soon it could be lifted and carried with the wind. The snow swirled and eddied, creating sculpted drifts over the landscape.

The following day was sunny and mostly clear, but a light wind blew and strengthened as the day wore on keeping a distinct chill in the air.

All spring-like activity had ceased even after the storm had blown eastward. Most of the birds had once more retreated to their former places of seclusion and the land lay quiet beneath the bright blue sky. Once more a man was reminded that it was still February.

In the woods a nuthatch laughed to itself, breaking the silence. A chickadee commented that the snow was "knee-deep" and another one near by readily agreed. One of the little chaps lingered awhile in some nearby sumacs to inspect some of the older, dried out seed bobs for hibernating insects. Once in awhile it would pluck one of last years' plump red seeds from its bob and, holding it firmly between tiny toes, extract all that it found edible.

Retracing his steps, a countryman notices that the pussy-willows are emerging just in time to greet the March Lion. Like tiny Angora kittens they cling to their branches, resplendent in their soft, silvery-gray coats. Nature's calendar is showing that it is but a few more weeks until Spring.

All day the wild geese moved northward. Flock after flock, wave after wave they came from the southlands. Formation was superimposed upon formation at times. Often a dozen or more flocks could be seen all at once. And there was an abundance of goose music. Their honkings filled the sky continually, and a countryman reveled in the almost supernal sound.

Although rain developed that evening along with light fog, the geese continued their northward flow until mid-morning the following day when their numbers diminished to a few scattered flocks up the lakes and in between.

Two whistling swans flew silently overhead just a few hundred feet from the ground, their great pinions propelling their ghostly forms to some distant destination to the northwest.

Great flocks of red-winged blackbirds and common grackles arrived the same day. Treetops filled with red-winged forms and their “o-ka-lee’s” rang out clearly over the countryside, hours on end.

Clouds and sun debated the right to rule the sky and an agreement was apparently reached to take an equal turn during this delightful day.

Robins bounced across lawns in search of newly-thawed worms that had recently come to the surface while yet others chirped saucily from roofs. House finches sang snatches of their spring songs and the dulcet tones of the bluebirds added to the symphony.

Five chickadees seemed to have acquired a dose of spring fever as they chased one another in and about a brookside clump of bushes, communing amongst themselves in a variety of notes and tones so common to their species. From a hedgerow came the sprightly song of a song sparrow, not the full song but abbreviated phrases, as though the bird was slightly out of practice.

All of Nature seems to realize that Spring is but a few weeks away and the tempo of their activity increases a little more as each day passes. Starlings are tuning up with their various notes and bits of mimicry and are beginning to forage the fields and lawns for whatever they can find.

Gray squirrels are becoming more sociable and pairs are often seen following each other up and down tree trunks, over and around limbs, all the time as though connected by a relatively short tether. Robins are beginning to act a trifle aggressively towards one another, more by their chirrups than the body language that will become more apparent as the season progresses. Nevertheless, there is an undeniable spring-like atmosphere and it has its effect upon the wild folk.

March is a fickle month; it always has been, and totally unreliable. It may have a mild beginning and a frigid ending or it can be wild and windy all month. One takes their chances with March and patiently waits for April. Meanwhile, a dusting of snow overnight is never unusual and squalls over the ridges and across the valleys are normal this time of year.

A countryman has learned to take what comes as a matter of course and this morning ramble ends amid snow flurries, sunshine and the song of the sparrow.

Despite the occasional snow and icy blasts from the Arctic, the wheel of the seasons continues to turn, inexorably, towards Spring. In a few more days it will be here, at least by the calendar but not necessarily by the ever-changing weather. Nevertheless, despite the weather, processes are going on that cannot be hindered. Buds swell, sap flows, birds migrate and animals emerge from hibernation even as snowflakes tumble and twirl earthward and may even temporarily cover the ground.

Mourning doves now coo plaintively on mild mornings and male robins have become combative concerning territorial rights. Geese continue to migrate northward but their flocks have become fewer and farther between. The grand migration has passed, another evidence of the advancing season.

A male marsh hawk, now known as a harrier, sails low over pastureland and field in search of small rodents. A turkey vulture floats buoyantly over the vineyard, its wings catching each whisper of a breeze. Both birds are early in their arrival but considering the mild weather of late it is not too surprising to a countryman to see these two old friends so early in the season.

In the cemetery, a man walks more slowly as he approaches a certain horse chestnut tree. He usually has seen a red squirrel in this tree when he has passed this way. At first there is no sign of activity, then suddenly a flash of dull red streaks into the open and across the limbs of adjacent trees, nimble little feet scarcely touching the branches in their passing. Reaching the hollow bole of an old apple tree, the little squirrel disappears from sight. In these few brief moments a man noticed that the little fellow was carrying what appeared to be a chestnut in its mouth. This no doubt would provide a meal at some later date. Just now, one little squirrel was waiting for his rapidly beating heart to slow down before he did much of anything.

Old catnip stalks lay flattened to the ground. At their base small tender new leaves are showing green and eagerly greet the morning sun. Many frosty mornings lie ahead and they wisely huddle amongst last year's fallen leaves, wisely measuring their growth with the warmth of the progressing season.

Homeward bound, a man listens to the voices of Spring, voices that are familiar yet have changed in tone and cadence with the lengthening days. Flickers and cardinals are in a courting mood and their calls and songs reflect the season. Titmice, chickadees, nuthatches and woodpeckers all voice their opinion that winter is over and that it is time to get on with spring-induced activities.

Moving through low bushes a man once again comes upon a polyphemus moth cocoon. It dangles loosely from a silken strand that is attached to a slender twig. This time the cocoon is heavy with life and a man is happy that this is so. He carefully notes its position and will patiently wait for the warmth of June to complete the transformation within from dormant pupa to emerging moth.

# SPRING

What would the month be without the March Lion making a grand entrance sometime during the thirty-one days allotted to it? Add to the leonine presence a visit from the Arctic Bear and all previous thoughts of an early spring quickly vanish. When those two rambunctious rascals get together almost anything can happen.

Bits of greenery, responding to an earlier warmth, suddenly become covered with snow or are nipped by temperatures that suddenly dip nearly into the single numbers.

The wild winds strum the boughs of pine, spruce and cedar eliciting vibrant chords peculiar to each species. Brown leaves vibrate in unison on the young white oaks with each passing gust of wind. Wind-driven snowflakes cling to and slowly build up on the leaves muting their vibrations.

The diverse arrangements of the branches of the maples and willows provide varying harmonics of their own sometimes adding notes of sadness as the wind produces laments from the tortured and twisted branches.

Finally, the forces of wind and weather pause for a time as if to catch their breath. This respite sometimes allows the sun to shine through thin clouds and shed a gentle warmth upon the land.

From the nearby woods a titmouse calls in a sprightly manner and is answered by another. A crimson cardinal alights atop a small walnut tree and, in cheerful notes, proclaims his gratitude for the cessation of wind and snow and the return of the sun's warming rays.

A man must not conclude his walk until he visits the chestnut tree where the red squirrel is often seen. Since his last visit a countryman has learned that the bitter fruit of the horse chestnut is too acrid even for squirrels. Apparently what the squirrel had previously carried away was some other nutritious tidbit.

The little squirrel is soon spied perched on a knot protruding from the trunk of the tree about a third of the way up. It is busily consuming something to its liking, turning the morsel dexteriously with dainty forepaws as it eats. Its snack finished, it scampers up the tree to a place where it can sit close to a convenient cavity. A breeze ruffles its tail briefly as a man retreats, and a man is content in knowing that the little squirrel is alive and well.

When a countryman considers March his thoughts immediately turn to red-wing voices and the furry forms of pussy willows. But an even more certain sign of approaching Spring is the arrival of the killdeers. The local pair arrived in time to leave their dainty footprints in the remnant of the last snow. This is not at all unusual, for as long as a man can remember the arrival of the first killdeers often coincides with a late winter or early spring snowstorm. They are resilient birds and seek out wet or damp areas in pasturel-ands, fields or around farm ponds.

A man will always savor the plaintive cries of the killdeers from the farm fields or from the sky overhead for these announce the soon demise of winter and the imminent arrival of Spring.

There's an old saying that goes something like this, "If you don't get the snow in the winter, you'll get it in the spring". This country saying was brought to mind the first full day of Spring when several inches of snow fell during the night and the storm did not seem inclined to let up any time soon.

For the most part, the wildlings retreated to their favored places of shelter to wait out the storm but the inclement weather didn't dampen the ardor of a male house finch who sang its courting song from the sheltering boughs of a large spruce tree. A cardinal's calls of "cheer, cheer, purty, purty, purty" rang out over the countryside and the dreary day was brightened considerably by his clear, clean notes. A bluebird bravely added his dulcet voice to those of the others before finally seeking shelter.

Several crows sat, somber sentinels of the ermine-clad countryside, in their respective tree-tops.

A few days after the storm things returned to normal. First it was the bugle-like calls of the bluejays and then the sprightly song of a song sparrow that broke the silence. Robins chirruped as they congregated along ditch lines and other wet places in search of worms. Already, these areas were becoming green as new blades of grass pushed upwards towards the sun's warming rays. Here and there a few robins started caroling as if in thanksgiving for sun and sustenance.

The limbs of the silver maples are covered with buds bursting at the seams. Other trees also sport swollen buds and emerging catkins.

Several juncos, congregated at road's edge, fly into a nearby cedar at a man's approach. Nearby come the tenuous notes of a white-crowned sparrow hidden in the depths of the thickets. Flickers practice their spring calls while the red-wings get down to the business of staking out their territories. A killdeer cries piteously as it flies erratically overhead, but its very presence betrays a bravery that supersedes any apparent sense of sorrow. Each year they arrive during stormy periods and accept the challenge.

A dark spot in a distant field draws a man's attention. Focusing his binocular upon the object, he discovers not one but three animated objects which turn out to be three wild turkeys practicing their up-coming courting rituals. Silhouetted against the snow, they take turns strutting about in tight formation, one so tight that at times there appear to be only two birds. To accomplish this, one bird would relax tail and wings thus diminishing its size and blending into the forms of the other two as they continued their display.

A countryman never ceases to enjoy watching the mating rituals of the wild turkeys. Their pompous displays and struttings are an added assurance that Spring is here no matter what the weather may be.

Unseasonable warmth has contributed to an early Spring in spite of the fact it was punctuated by a snowstorm. March came in like a lion wearing an overcoat and boots and departed carrying them over his shoulder.

Spring peepers climbed out of their muddy beds and assembled choruses all over the countryside. By the end of the month a few toads had roused themselves enough to add their trills to the chorus, but their voices lacked the energy that will be forthcoming in another month.

With temperatures climbing into the seventies and eighties for about a week, leaf buds on many bushes and shrubs released new little leaves. These were followed by the black willows and other small trees and soon a pale green haze appeared in patches across the landscape.

Everywhere one looks now there is green grass where a short time ago there was only tans and browns. The renewal of life has begun once more.

Flies emerge from their hibernation, small insects hover in thin groups while larger ones emerge en masse to plaster themselves across the windshields of fast moving vehicles.

Phoebes have arrived, early by the calendar, but in time to feast on the abundant insect population. A countryman watches each day now for the arrival of the first tree swallows. Barn swallows and their cousins will arrive a short time later. The emergence of the insects and the arrival of insectivorous birds is timed quite precisely each year and is part of the great plan of things.

Wild turkeys are on the move and toms are increasingly active in their courtship displays as the time for that serious pursuit rapidly approaches and the hens become more receptive to their attention. A flock foraged in a field with four deer early one warm morning creating a congenial wildlife pageant.

A long-legged blue heron flew towards the lake one morning. The wind was just right so that it could set its wings now and then in a downward arc and sail for quite a ways before having to resume its flight.

A pair of horned larks searched for food along a roadside for some length of time before making their departure. Horned larks are brown ground birds larger than a sparrow. They sport black "sideburns" and two small, feathered "horns" on the top of their heads, giving them their name. Also, like all true larks, their hind claws are very long and almost straight. They are often seen in flocks feeding on seeds and insects.

From fallow fields meadowlarks call out "to see you, to see yer". The flash of white outer tail feathers and their distinctive flight pattern of several rapid wingbeats alternated with short glides combine to assure a quick identification of these residents of fields and meadowlands.

Regardless of changeable weather patterns a countryman knows springtime has arrived when he hears the song of the field sparrow coming from the vineyard. He knows winter is vanquished when he hears the trills of the juncos from their concealments in the roadside bushes. This fact is made even more evident when woodchucks start their spring house-cleaning and debris and fresh earth appears at the entrance to their burrows.

A pair of Canada geese makes their way towards the lake in the foggy early morning sky. They converse between themselves as all proper mates should, perhaps discussing the day's activities or possibly where they intend to have breakfast.

The robins are pairing up and at least one female has already decided where she is going to have her nest, while her mate defends their territory.

The chipping sparrows are back and add their trills to the carols of the robins creating a pleasing atmosphere to April mornings.

Song sparrows busy themselves around buildings and garden areas and out in the hedgerows, stopping their activities now and then to add their sweet songs to those of the others. Field sparrows pleasantly proclaim possession of vineyard and field, their voices floating across the countryside.

A flicker calls from the woodlot and a female hurries in answer to his call. A downy woodpecker drums on dry dead wood in hopes that his tattoo will also be answered as was the flicker's calls.

Each day more tiny leaves are unfolding on various trees and shrubs, adding to the variety of pastel shades of green. Darker shades line the watercourses and swamp-edges where grasses grow rank and dark green.

Catkins dangle from the branches of poplars and willows while the maple branches are decorated with colorful flower buds bursting at their seams. New leaves have emerged along the tall bramble canes and in a few more weeks will be followed by bursting flower buds.

Red-winged blackbirds perch about the edges of their territories, patiently awaiting the arrival of their mates and displaying their colorful wing epaulets each time they break forth into their gurgling songs.

Bluebirds inspect birdhouses; first one then the other of a pair will enter the house, linger inside a minute or so then emerge to perch on the sloping roof and warble softly to its mate. Obtaining and retaining possession of a birdhouse is often a difficult job for the bluebirds. House sparrows are very aggressive and tend to dominate the more gentle bluebirds, either gaining possession first of a house or driving the bluebirds out when they start nest building.

A pair of mallard ducks wing swiftly across the late afternoon sky, quacking noisily back and forth. A great blue heron flies slowly homeward on broad wings, its long legs trailing out beyond its stubby tail.

Shadows lengthen and robins commence their evening vespers and continue for a short time even as dusk deepens, covering the land with its dark velvet cloak. Finally, all bird voices are silent leaving the peepers to fill the night air with their clamorous voices.

The cardinal's clear notes rang out on the morning air. Perched atop a scrawny tree, its bright red suit almost glowed in the sunlight. Repeatedly it called and like an echo it was answered by another cardinal not too far away.

The mockingbird also was in fine fettle, going through its repertoire over and over with variations here and there. Mixed in with all the mimicry come the perfect notes of a phoebe immediately followed by other less intelligible yet wonderful tones and notes that only a mockingbird can produce.

From the depths of a thicket came the tenuous tones of a white-throated sparrow. Only briefly did it sing, then all was quiet except for the mocker in the distance.

A huge flock of cedar waxwings congregated in the top of a large black willow, their numbers spilling over into neighboring trees. A man counted well over forty of them and then lost count because of the constant movement at the outer edges of the flock. Several small flocks descended to feed on bright red berries on the bushes below and a short distance away. A man thinks that cedar waxwings are among the best-dressed birds in their smartly tailored suits of tan and fawn, decorated with touches of yellow and small spots of red.

After a night of rain, colors are subdued beneath leaden skies and droplets of water hang from the branches. Small footprints in the mud along the stream edge tell of the passing of a raccoon during the night.

Overhead, a vulture sailed low, just skimming the tree-tops, its wings catching every whisper of a breeze. Its passing attracted undue attention from a pair of crows as it entered what they considered their private domain. Paying the crows scant attention it sailed serenely on its way.

Catching sight of a countryman's approach, several turkeys walked briskly across a fallow field where they had been foraging. Earlier on a man had noticed what he had been sure were fresh turkey droppings so he had kept on the lookout for them and apparently saw them at the same time that they had observed his approach.

Thin, sweet notes from a chickadee and the laughter of a nuthatch come from the woods and a titmouse calls anxiously for Peter. Redwings gurgle from atop their respective shrubs across a field.

Without the aid of direct sunlight, cherry leaves unfurl and a scattering of blossoms reveal their creamy whiteness in contrast to the otherwise drab day. Soon they will engulf the bare branches with their blossomings and warm bright days will bring bees and other insects to partake of nectar and spread pollen from tree to tree, ensuring another crop of cherries.

A countryman reflects that despite rain-washed nights and drab days Spring moves ever forward, into blossom-time and on into days of sunshine and fruitfulness.

Chokecherry blossoms brighten the hedgerows and woodland edges. Wasps and bumblebees are becoming more active with each passing day. A dragonfly hovers, moves sideways then up several feet before darting away out of a man's sight. Myriads of small insects have hatched and will provide sustenance for the swallows and flycatchers as they move northward.

Quaking aspen leaves quiver in the gentle morning breeze and shimmer in the sunlight. Already, some of the cherry blossoms have fallen to the ground, loosened by April showers and their accompanying winds.

Sparrows sing from every quarter it seems. The voices of chipping, field and song sparrows fall on a countryman's ear, each from its own particular environment.

A female cardinal attracts a man's attention as she unobtrusively moves among the low shrubs and bushes. Could it be that she is looking for a good place to build her home? As a man watches, the male sings clearly a few yards away, attracting attention away from his mate and proclaiming his territory. She moves farther away and he soon follows, but at a distance and continues singing.

The mockingbird speaks distinctly from a brushlot while a meadowlark cheerily calls from a nearby field. A redwing sings and waits for a mate. A jay shrieks shrilly from a stand of pines while crows court close by in a garrulous sort of way.

Green has now become the predominate color across the countryside. Wheatfields, fallow fields, hayfields and pasturelands are all a living, vibrant green. A light green haze has spread across the upland slopes and down the ridges between the lakes.

The winds pick up and several red-tailed hawks soar in ever-widening circles as they watch the ground below for possible food.

The earth warms and a vulture catches the rising thermals on outspread wings. It sails over the countryside, wings tilting now and then in response to the varying air currents. Momentarily it drops lower to the ground, then lifts effortlessly upwards as it approaches a woodlot and is soon lost to sight.

Bright yellow dandelions march up and down the green grass carpet that lies between the vineyard rows. A woodchuck dines on the bountiful supply of spring greens out on a headland.

Buds and blossoms, birds and bees are increasing in numbers just as the sun climbs steadily higher into the sky as the month draws to a close. April has once more fulfilled her obligations for another year.

May could well be called the month of blossoms. Of course it started a month or so ago when the dandelion-like blooms of colt's-foot brightened the roadside ditches. Even before that the swamplands had their skunk cabbage. With colt's-foot the hepaticas made their appearance, emerging from their beds of rich, leafmold soils of open woods and forest slopes. Their colors varied from lilac to bluish to white. As a

man's mouth waters in anticipation but there remains about six weeks between anticipation and full realization.

Black flies have emerged and buzz hungrily about a countryman's face. One alights on a bare arm and manages to gain a bit of sanguine nourishment before its need is suddenly eliminated. A man is glad that they are nowhere near as prolific as they are in the north country where they can make life nearly unbearable from the first of May until after the first fall frosts.

A killdeer cries plaintively from overhead and is answered by its mate. Tree swallows pick up bits of straw and dried grasses from the edge of the road and fly to distant nest boxes. A robin is busily engaged in coaxing angleworms from the ground and carrying them to a nest already containing small robins.

Lilac blossoms scent the spring air with their pervading perfume. Apple orchards are overspread with a blanket of white blossoms and as one draws closer the hum of bees can be heard as they busily go about their business of gathering nectar and spreading pollen from tree to tree.

Clouds build as the afternoon warms and soon their forms merge into a gray cloak that covers the sun's glowing sphere. The temperature drops several degrees in response and scattered drops of rain cause a man to haste his steps towards home. Pink-petaled wild geraniums greet a man from the grasses as he steps along. Moving along a woodland edge he notices the flourishing forms of mandrakes or May Apples. Their blossom buds droop pendulously from the elongated stems.

Thunder grumbles ominously close by and the rain drops thicken as a man barely reaches shelter before the afternoon thundershower sweeps across the countryside, cleansing and sweetening the atmosphere.

Small green cherries have replaced the white blossoms that so recently cloaked the choke-cherry trees while the blossom buds on the black cherries are just nicely becoming receptive to the probings of the bees.

A yellow warbler intently inspects the limp new black walnut leaves for insects and at the same time does not neglect the equally flaccid catkins that dangle from every branchlet.

An oriole greets the morning sun with his clear, lively song, a song that a countryman has waited all winter to hear once again. The barn swallow, the catbird and the oriole all are an assurance that spring is here and that summer cannot be far behind.

Boblink voices bubble joyfully from meadowlands and fallow fields as swallows skim low overhead against the azure sky.

As the tree swallows a week ago were busily gathering bits of straw and grass stems for their nests so now the barn swallows can be seen at the edge of the farm pond selecting mud of the right consistency with which to build their nests against barn beams or the roof rafters of outbuildings and sometimes even porches.

A bright-eyed baby cotton-tail chews contentedly on a dandelion leaf, the afternoon sun warming his soft brown coat. A little later he hops along a little pathway, kicking up his heels in the exuberance of youth A countryman wonders how long he will survive the harsh realities of the wide world around him.

Ants have rebuilt their earthen mounds, repairing the ravages of winter past when they were lying dormant well below the frost line. Now the ant hills are alive, their tenants actively living up to their reputation for industriousness.

Dandelions nod their now-hoary heads in the late afternoon breeze. Their once golden blossoms have matured into a vast multitude of tiny seeds, each with its own flossy parachute ready to carry it to some distant place where a new "lion" can go on the prowl.

The earth continues to turn, drawing the day to a close. The sun's last rays paint the thin western clouds crimson and robins commence singing their evening carols. Tree frogs call their welcome to the twilight, tentatively at first then with increasing enthusiasm.

As shadows lengthen and dissipate, a wood thrush sings from the near-by woods. Its melodious voice adds sweetness to the fresh evening air. It sings for several minutes and then retires as dusk envelopes the surrounding countryside.

Toads trill from their seats around the farm pond and along its outlet and a few peepers call shrilly in the dark distance.

A countryman senses the nearness of summer as May draws to a close. It is nearly a month away, yet everything in the natural world is heralding its arrival. The trees are all now clothed in their verdant beauty, their new leaves ready to hide bird homes, wasp nests and provide necessary concealment for the leafy summer squirrel homes.

These same leaves make it more difficult for a person to see the animals and birds, but what one cannot see he can still hear. A crested flycatcher calls "whee-eep"! A man scans bare branches, branch ends and even tree tops but to no avail, but he knows that the flycatcher is someplace nearby. From the leafy undergrowth a chewink calls "sreee!" Again a man peers into the verdure but not even the usual kicking over of dried leaves is heard, just the seemingly disembodied voice calling "sreee"!

Yellowthroats that usually peer out from the undergrowth through their black masks remain hidden, calling "witchity-witchity-witchity-witch", much to a countryman's frustration. Upon reflection however, he is grateful to be able to hear and to recognize the owners of the varied voices that he hears. As he listens to their songs he can see, in his mind's eye, the forms and colors of their owners and this is sufficient.

Ah, at last! A male cardinal is a flash of color as it flies across a clearing and a hummingbird makes its rounds through a patch of white and lavender phlox, hovering and darting about the colorful blooms. A yellow warbler sings nearby and a countryman catches a brief glimpse of lemon yellow among the fresh new leaves of a shagbark hickory.

As a man heads towards home he notices the blossoms of buttercups, shining in the sun like so many newly-minted golden coins. Then he spies some butter-and-eggs by the wayside, their newly-opened blooms freshly painted yellow and orange. Black cherry blossoms scent the spring air with their delicate fragrance, their drooping flower clusters soon will be replaced with small green fruits just as their cousins the chokecherries did a few weeks earlier.

A speckle-breasted young robin hops slowly across the lawn as a man approaches his doorstep. Its father protests loudly at his approach and the youngster flies around the corner of the building.

Young birds and animals and late spring flowers are all sure signs that summer is patiently waiting just off stage.

May emerges effortlessly into June and small, green May Apples have formed where bright white blossoms once peeked out from beneath the umbrella-like leaves above them.

White blossoms cloak the bramble canes and bumblebees and honey bees buzz and hum their way among the flowers.

Hornets and yellow jackets and the big, bald-faced wasps are hard at work transforming dead wood into paper pulp suitable for the construction of their varied types of paper homes.

Mud dauber wasps gather globules of mud from puddle edges, their steel-blue bodies glistening in the sun, their wings flicking nervously. One leaves and soon another arrives to take its place. The mud homes are made up of several compartments, each containing an egg and a small spider upon which the emerging larva will feed. The spider has been paralyzed so that it will be fresh when needed. Upon reaching maturity, the new wasps chew their way through the mud walls of their home to freedom.

At the blackberry patch a white-crowned sparrow alights sideways on one of the thick canes. A countryman watches it for several moments until a song sparrow suddenly appears and sends the white-crown fleeing into the vineyard. A man is not surprised at such a turn of events, for as long as he can remember a pair of song sparrows has called the thorny recesses of that patch home each year. They scold him as he picks the big juicy berries and each Spring a man listens to the male proclaim the patch of briars his private domain. Thus it came as no surprise when the white-crown was made to feel less than welcome when it inadvertently trespassed thereabouts.

Overhead, three crows make life miserable for a hunting red-tailed hawk. The hawk squeals petulantly, executing several barrel-rolls and presenting deadly talons to its tormentors. In the end the hawk reluctantly leaves the area to hunt elsewhere.

For the most part the peepers and toads have dispersed over the countryside for the warmer months, leaving behind in the ponds and pools masses and strings of jelly-covered eggs to hatch into wriggling tadpoles. Before summer's end, they will have grown and transformed into miniature replicas of their parents.

The calls of the peepers and the trills of the toads are seldom if ever heard now, but as the shadows lengthen in late afternoons the bird-like calls of the gray tree frogs can be heard. The calls become more numerous and increase in tempo as darkness spreads its mantle over the land and another day draws to a close.

In the woods-edge near the cemetery the red squirrel scolded. Even though a countryman couldn't see it, knowing red squirrels as well as he does, he could see in his mind's eye the little hind feet stamping on a limb, red tail jerking spasmodically, accentuating the hiccup-like vocalizations. These were followed by rather shrill chirring calls that undulated with their utterance.

Two crested flycatchers took turns sallying forth from their dead branch lookouts to snatch insects from the air. Their calls of "wheeeep!" rang out over the countryside. These medium-sized birds can be recognized not only by their whistled calls but also by their cinnamon tails and wings and yellowish bellies. Their airborne assaults upon flying insects also identifies them as cousins to the phoebe and the kingbird.

A chipmunk busily gathers maple seeds, neatly clipping off the propeller-like wings before stuffing the seeds into its ever-expanding cheek pouches. These pouches are very elastic and are fur-lined to keep their contents dry as well as secure. When no longer able to hold anything more, the pouches give the chipmunk the appearance of having a very bad case of mumps.

The local mockingbird fills the country air with a combination of musical calls, notes and other vocalizations. Sometimes he may sound like several songbirds all wrapped up into one, then again he may mimic one to such perfection that for a moment or two he can even fool a countryman until he changes tone and pace once more. He likes to perch atop a tree or dead snag so that the whole world can see as well as hear what a great songster he is. There is nothing bashful about a mockingbird.

As the month of June matures so also do the wild strawberries, quickly ripening in the late spring sun. This year, for some reason, they do not seem to be as prolific as recent years despite the earlier abundance of blossoms. Perhaps it is due to insufficient cross-pollination or something about the weather. At any rate a man enjoys the flavorful few that he discovers during his rambles afield.

Walking across a fallow field towards home, a man notices movement in the grass near his feet. Since one never knows what may be hidden in the grass he bends cautiously over for a closer look. There, hopping awkwardly against the stiff stems of grass, is a large plump toad. A toad this far afield is a clear indication to a countryman that the Wheel of The Seasons keeps turning, bringing with its turning the growing and the maturing of early summer.

The world is lush and green; everywhere it is overflowing with new life. Cute little cottontails and small woodchucks allow a countryman to approach them much too closely. Although these youngsters are enjoyable to watch in their natural environment, they can quickly devastate portions of country and suburban gardens. Sometimes it is difficult living with wild neighbors during the growing season when one's food supply appears, to the wildlings, an easy source of tender, succulent food. Despite these conflicts, a countryman would not want to live where there was not wildlife. They are an integral part of the world we live in and their absence would not only make our lives less enjoyable but, in the long run, create other problems that would far outweigh the problems that they sometimes create by their presence. Sometimes when the young beans, carrots and Swiss chard have been gnawed to ground level this is hard to accept.

Mid-June finds the oriole singing in the tree-tops, foraging in the orchard and helping raise his family in their closely woven bag-like nest swinging at the tip end of a slender branch.

Indigo buntings sing sweetly along field and vineyard edges. Since they are birds of the edges, they are easily victimized by the parasitic cowbirds. No doubt that is one of the reasons their numbers have diminished in recent years. More and more of our songbirds are raising cowbirds instead of their own young which the young cowbirds push out of the nest to eliminate competition for food.

Red-winged blackbirds and bobolinks are busily raising their families in the meadowlands and fallow fields. Jays and catbirds are also busy and are more quiet now so as not to reveal the location of their nests.

Field daisies brighten the landscape with their golden-centered white blossoms while wild roses blush a deep pink along country roads. Orange and yellow hawkweed or paintbrush splash their bright hues across the canvas of the fields.

Reflecting upon all of these things, a countryman knows that the season of growth must soon merge with the season of maturity and fruition.

# SUMMER

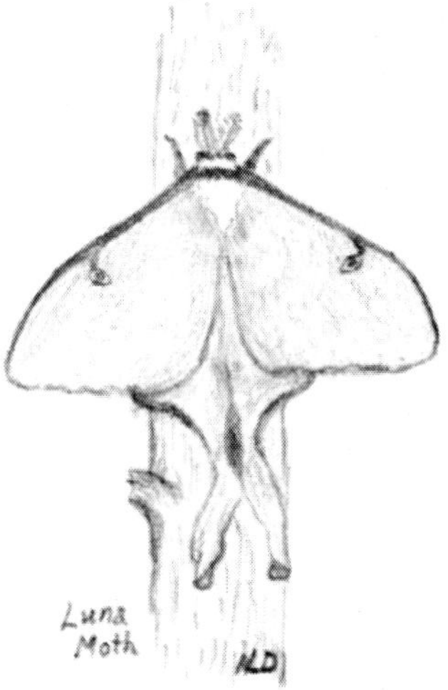


The Summer Solstice has come and gone for another year. Already the days are diminishing in their length, but at this point it is imperceptible except to the astronomer and his instruments or to someone who is willing to carefully note the exact time of sunrise and sunset. For all practical purposes the long days and short nights of summer will continue for several more weeks.

Large discs of creamy blossoms adorn the elderberry bushes while green berries have replaced the blossoms along thorny blackberry canes. Black raspberries are daily deepening in color and a countryman waits patiently for them to become black and sweet.

Chokecherries are also showing color and before long they will provide nourishment for a variety of songbirds and small mammals.

Swallows fly over the fields and vineyards and pasturelands in pursuit of insects with which to feed their growing fledglings.

Occasionally a butterfly or two may be seen fluttering about early summer flowers. A few yellow tiger swallowtails, a white admiral or banded purple and an occasional monarch are some of the butterflies that a man has observed of late. Cabbage butterflies and a few sulphurs are seen more often.

With the opening of the milkweed blossoms monarch butterflies and others will be seen hovering about the sweet-scented blooms.

Wild morning glories lift their small pink and lavender trumpets to the sun, bringing color to the vineyard rows and roadsides.

Buck deer are growing new antlers beneath their protective coverings of velvet. Spotted fawns gambol in the sunshine and follow their mothers obediently. Here and there can be found signs of their foragings, young shoots of elderberry and other tender plants chewed off several feet above the ground.

As the sun nears the western horizon, several chimney swifts course across the sky, chattering to one another, their wings "twinkling" as they quickly pursue their evening meal. A man watches admiringly their deft maneuvers and swift flight for a few moments and then makes his way homeward as shadows lengthen across the land and green frogs croak from the nearby pond.

The long days of summer have finally arrived; hot, hazy and humid at times, yet filled with so many good things that the heat and humidity are made more tolerable. After all, heat and humidity are needful for the growth of all plant life.

Fields of new-mown hay fill the air with a special fragrance, the fragrance of early summer, the last days of June and the early days of July.

Bobolinks and red-winged blackbirds are busily rearing their young in those fields not yet touched by the mower. Their voices too are the melodies of summer and are a pleasure to listen to.

A towhee calls from the tangles of a brushy area. They often are seldom seen except when they fly, low to the ground, across an opening. To a countryman, a brushlot or a hedgerow is not complete without at least one towhee and its cheery calls of "tow-hee, tow-hee"!

Barn swallows twitter companionably as they fly to and from their mud nests in barns or beneath the overhangs of eaves.

Sometimes the clear calls of a crested flycatcher cut through the heavy air of a summer afternoon and the husky voices of chickadees come from the edge of the woods where they are feeding their fledglings.

Each day changes can be seen being made. Great mullein stalks grow steadily taller, their tips adorned by lemon yellow florets that give to them a glow of gold.

Green blackberries swell with each passing day while black raspberries turn an ever deepening red. Wild strawberries have had their days in the sun and are now a thing of the past for another year.

Blue-eyed grass, with their dainty flowers of blues and lavenders, peer up at a countryman when he goes for a closer look at a clump of black-eyed Susans.

A man knows summer has arrived when black-eyed Susans combine their bright colors with those of daisies and red clover along country roads. He accepts the presence of summer also when he finds a tree of sweet red cherries in an out-of-the-way place and eats his fill before turning his steps towards home.

When one thinks of July one thinks of hot and humid days often followed by warm and sometimes equally humid nights. One thinks of fireflies and mosquitoes on sultry summer evenings and bats swooping after their insect supper or no doubt, to them, it is their breakfast after sleeping all day. July is a time for fireworks and nature provides her own in the form of brilliant lightning flashes followed by window-rattling crashes of thunder. The accompanying rains serve to refresh the land and, with the summer sun, perpetuate the continual growth of things until maturity is finally reached.

A countryman will now take a dish and wander afield to favored patches of black raspberries. As the dish slowly fills, he listens to the chatter of a house wren and perhaps also the clear calls of a cardinal. He may see a chipping sparrow carry a small green caterpillar to its young hidden away in the nest among the thick boughs of a spruce. This same spruce has small green cones glistening with droplets of pitch, hanging from its branchlets.

Dark, sweet berries finally fill the dish to the rim and a man starts towards home. Crows caw lazily from the woods beyond a cornfield and a chipmunk chucks monotonously from atop a rockpile. A dozen feet or so from a vineyard row a woodchuck hungrily feeds on succulent plants and an occasional chicory leaf. As a man approaches, the woodchuck sits up warily, a chicory leaf drooping from its now motionless mouth. Several moments pass and then it suddenly drops to all fours and runs to the security of its nearby burrow.

Yellow and black goldfinches inspect tall thistles but find only wine-colored blossoms. It will be more than another month before seeds will have formed and thistle down will become available to line their dainty cup-like nests. They soon take wing and their calls of per-chicory, per-chicory" drifts across the land.

A damp area yields one small hoof print of a fawn, the only sign of its passing or even of its existence for that matter.

As a man nears home he notices a bluebird family searching for bugs and other insects among the tops of a vineyard row. He pauses a few moments and is rewarded with the sound of the parents' soft, sweet warbles as they converse with their young ones. Berries and bluebirds have enriched his day and he contentedly continues on his way.

A countryman has a robin for an alarm clock, albeit a rather early one. This particular robin starts caroling around 4:20 each morning and continues for over half an hour. Most mornings a man is sleeping soundly at that time, but if perchance he should become somewhat awake he listens to the joyous greeting of the new day until at last he is lulled back into peaceful slumber.

Speaking of robins, several fully fledged yet still speckle-breasted youngsters were seen with their elders in the vineyard. They appeared to be finding abundant insect fare much to their liking as they were active up and down the rows.

Yarrow plants have now grown to the point of conspicuousness and once again a man is able to enjoy the scent of one of their crushed leaves as he has done each year for many years. It provides one of the many enjoyable scents of summer to one who appreciates the changing seasons.

A hummingbird darts across a man's path and the flash of white against gray betrays the presence of a mockingbird. Somewhere within a thicket a catbird mews several times and then all is quiet.

The day warms and thermals rise from the heated ground providing air currents for the four turkey vultures that come into sight. Keeping a generous distance from one another, they tilt and glide, dip and soar effortlessly on outstretched wings.

Two young gray squirrels eat green maple seeds while close at hand a young cottontail munches on a white clover. The young of both species feed harmoniously until their appetites have become satisfied.

The day draws to a close and a blue heron flaps laboriously homeward on broad wings, its neck tucked neatly into its shoulders and long legs extending beyond its short tail. Green frogs croak from the farm pond.

The sun sinks slowly beyond the western ridges, its last rays gild the scattered clouds, creating a glorious golden sunset that is wondrous to behold. Once again, the Master Artist has created a masterpiece across the canvas of the sky.

Hot and humid days make the blackberries ripen, especially those which grow where the sun can reach them during much of the day. Those in more sheltered or secluded areas ripen much more slowly and provide a harvest later into the month of August.

A countryman already is able to pick small dishes of the sweet, dark berries. These go very well with his morning cereal during July and into August. The hot days are not suitable for a countryman to ramble about the countryside. They are not suitable for many of the wild folk either. Cool glens and ravines with a bit of running water in their depths are the place where one will be more apt to find birds and animals, especially during the heat of the day.

Some of the songbirds are busy with their second brood of young while others have already completed that task for the year.

Crows have been moulting their feathers for several weeks now. Almost everywhere a countryman goes he sees black feathers from wing or tail.

Soon the songbirds that migrate will follow suit, followed next by the year-around residents.

A man finds the period of twilight a congenial time of day in which to meander about the countryside. There is still enough light for observation and the daytime temperatures have moderated to a more comfortable level.

Passing an old brushpile, a man hears the notes of a song sparrow as it prepares to retire for the night. A cardinal sings a few clear notes from a secluded spot among some Virginia creeper vines before falling silent as the shadows darken. The same shadows do not dampen the spirits of a wood thrush who continues its melodic calls from the nearby woods.

A cricket chirps cheerfully along the pathway and as the dusk of evening deepens fireflies light their little lanterns and are soon twinkling all about a man as he carefully makes his way towards home.

As July draws to a close one comes to realize that the year is now reaching into the depths of summer. Hot, hazy days linger over the land and are followed by warm and often humid nights. Then, unexpectedly, may come cooler days with azure skies replete with cotton-candy clouds and a gentle breeze.

Black-eyed Susans brighten fallow fields and blackberries ripen rapidly in the summer sun. And deer flies bite those who venture near the cool woodland edges.

The flowers of late spring are being replaced by those of a mature summer. Bright blue blossoms adorn the scraggly stalks of chicory along the road edges. Farther afield, the spiny cones of teasel wear belts of lavender florets. Burdocks and bull thistles sport blooms of varying shades of purple. Saucer-shaped white flowers of Queen-Annes's lace nod drowsily in the afternoon breeze while goldfinches cavort over field and vineyard.

Grapes hang from their vines like so many green marbles. It doesn't seem possible, to look at them now, that within three more months they may well be processed into juice and jelly or into varieties of wine.

The yellow and orange spurred flowers of jewel-weed grow luxuriantly in damp, shady areas where they are visited by hungry hummingbirds.

Catbird parents scold a countryman when he inadvertently passes too close to a youngster on the ground. It flutters deeper into the undergrowth and is lost to sight and a man moves quickly away so as to calm the agitated parents.

Many young animals are seen as a man meanders about the countryside. A spotted fawn with its mother disappear into a brushlot, while young woodchucks and cotton-tails often allow him to approach much too closely for their own good. Young gray squirrels frolic in youthful abandonment across lawns and woodland floor, often forgetful of the dangers that ever lurk nearby. Some will pay the ultimate price for their forgetfulness yet, at the same time, will provide needed nourishment for the predator or its own growing family.

With the turning of the final pages of the month comes the assurance that more adventures await a countryman as he rambles on into the depths of Summer.

The afternoon is sultry and a bluish haze blurs the distant ridges. The dog days of summer have begun. A young robin flutters and chirps for parental attention from its shady retreat among huge basswood leaves while the parent birds forage in the nearby vineyard.

Elongated seed pods now are visible on sturdy milkweed stalks where, but a short time ago, there were fragrant pink blossoms. A countryman inspects the leaves of numerous plants for signs of monarch butterfly larvae but to no avail. This year, like the previous one, lacks the lively activity of the tiger-striped caterpillars as they ate their way to maturity.

A man remembers how, many years past, he would wander among milkweed plants and observe several caterpillars on nearly every plant. Sometimes he would collect two or three, along with some leaves, to take home and place in a jar to watch as they reached maturity, entered the pupal stage and eventually became beautiful monarch butterflies that would join legions more in their autumnal migrations southward,.

Some early autumn days the air would seem to be full of fluttering orange butterflies, the harvest of many milkweed patches. Those days have disappeared much as has the over-wintering habitat of the monarchs in many places in Mexico due to lumbering activities.

On a previous day a man remembers seeing one lone monarch fluttering along a roadside and he wondered then what its chances of survival were and whether or not it had managed to perpetuate the species,

A female cardinal flies up from a willow-lined drainage ditch and perches on one of the lower limbs. She shakes herself and droplets of water fly in every direction. She has just finished her bath in one of the few shallow pools below and now preens and fluffs her feathers repeatedly to aid in the drying process, As a man draws closer, a catbird flies up; it too has been bathing but manages to fly the short distance to sheltering shrubs where it will also re-arrange its feathers.

A female downy woodpecker hitches her way up a willow trunk, searching the rough bark for insects, larvae and eggs. A family of six chickadees gathers overhead and call back and forth in cheerful notes as they inspect a man at close range.

A man always enjoys the company of chickadees whatever the time of year, and this sultry summer afternoon is no exception. His only regret is that he has no sunflower seeds to offer the little black-capped birds.

When goldfinches commence singing their courting and territorial songs, a countryman knows that the last half of Summer is about to begin. Their vacation time over, the wild canaries start thinking seriously about nest building and the raising of families. Other signs in nature indicate also that it is that time of year.

Many of the thistles have gone to seed, providing food and soft thistle-down with which to line their cup-shaped nests. These are hung from the crotches of tree branches, quite often those of saplings. Many other seed-bearing plants will soon reach maturity, providing additional food supplies for adults and youngsters alike.

August is typically the month of scanty rainfall, resulting in hot, hazy days. Some years this is the time when one will hear the rasping calls of periodical cicadas against the background of shimmering waves of heat.

Already some grain fields are being harvested, while others need a bit longer to reach full maturity.

Flickers call "clee-up!" from hedgerows and woodlots. Occasionally the shrill voices of jays may be heard as family groups roam about the countryside. Wood pewees call from the deeper woods and from those same woods can be heard the voice of the ovenbird calling "teacher, teacher, TEACHER, TEACH" rapidly in crescendo.

The ovenbird is a member of the warbler family but resembles a small thrush, olive-brown above but striped rather than spotted below. At close range an orange patch can be seen on the crown of the head. It is usually seen walking on pale pinkish legs across the forest floor, occasionally flying a few feet higher to a low-hanging branch.

Its dome-shaped nest resembles an old fashioned oven and is concealed among the litter comprising the forest floor. This odd-shaped nest gives the little warbler its distinctive name.

Banded purple butterflies are sometimes seen about the willows which are the primary food plant for their larvae. An occasional monarch butterfly may be seen fluttering about the milkweeds.

The larvae of numerous butterflies and moths are now reaching maturity. Walking along woodland trails, a countryman will sometimes encounter a small green caterpillar, dangling from a nearly invisible thread of silk attached to a branchlet many feet above.

A man pauses a few moments to observe the tiny creature and then continues on his way to see what else the eighth month may hold in store for him.

An evening ramble in mid-August yields its own special pleasures. The heat of the day has dissipated yet there remains a warmth that encourages the insect orchestra to tune up for their evening performance.

Crickets chirp steadily along the wayside as a man passes and katydids proclaim their identity from the seclusion of the willow trees. An unknown entity sends its notes out from the depths of the vineyard and soon the orchestra commences its composition.

A field sparrow sings its vespers as long shadows deepen across the land and a night breeze dances among the trees in the hedgerow setting their leaves to whispering in cadence the gossip of the day.

A robin calls sharply and is answered by the liquid notes of a wood thrush from the darkened woodlot.

The darkness envelopes the blue haze that veiled the distant ridges and wisps and tendrils of mist rise from the meadowlands, forming and reforming as it rises from the heated earth and into the cooler night air.

A tree frog sends forth several bird-like calls then falls silent, leaving the night to the sounds of katydid and cricket and to that unknown entity that persistently calls from the depths of the vineyard.

Thin clouds glide across the full face of the moon and for one brief moment the outline of a bat is silhouetted against the yellow lunar disc. A few stars shine brightly enough to penetrate the thin clouds and they seem to wink at a countryman as he makes his way homeward.

A deer snorts and leaps from a headland where it has been feeding and runs swiftly between two vineyard rows. The faint musk of skunk adds to the sights and scents of the mid-summer evening.

Strawberries have been gone for nearly two months and raspberries have been absent over half as long. Now the blackberries are on the wane. Only a few are left in the earlier patches but on the north sides of hill slopes and woodland edges there may yet be a harvest for a week or so.

A countryman remembers back, many years ago, sharing thumb-sized juicy blackberries with a barrel-sized, foraging black bear in a backwoods bramble patch replete with finger-thick canes that towered well over a man's head. Within a few hours two twelve quart galvanized pails were nearly filled to their brims and a man and his wife would start the three mile trek towards home. Those berries made mouth-watering pies, thick jam for school and work sandwiches and dozens of jars of sauce to be relished on many a sub-zero Maine morning.

A man can remember further back, as a lad in the hill-country of north-central Pennsylvania, of carrying a peanut butter pail and helping his mother pick the same black fruit along sunny hedgerows and fallow fields and in the shadows of forest edges where the berries grew fat and sweet. He can remember the cicada's song droning on and on in the heavy August heat and the tormenting bites of mosquitoes and deer flies. He can still hear the plunk of the first few berries as they dropped to the bottom of the tin pail and then became more silent as they neared the top. He remembers too the quiet chatter between mother and son as they discussed the present and sometimes the past as well as the future, forgetting for awhile the cares and perplexities of everyday life.

A man stops his reminiscing long enough to pluck a few straggling berries from their nearly barren canes. No need today for a tin bucket or even for a plastic dish. The only fruit left are those being picked and a few bruised and broken ones left by the birds and a scant few more that need to ripen in the next few days.

The sky is a dull bluish-gray and the land is parched for lack of rain, but none seems in the offing. The roadway home is dry and dusty and the afternoon sun beats hotly down upon a man's back as he trudges along. A song sparrow chirps dispiritedly and a dust-colored grasshopper flies up from the track ahead. A certain listlessness seems to pervade the very atmosphere as puffs of dust curl around a man's plodding feet. But then, this is late August and it's supposed to be hot and dry and grasshoppers are supposed to fly up on crackly, whirring wings from dusty roadways.

Summer matures into September and all the promises of May are being fulfilled. Once again the countryside is becoming filled with gold. While the flowering flames of the great mullein plants are flickering near the tops of their stalks, they have already been replaced with beautiful evening primroses whose golden blooms attract sphinx moths which hover about them as dusk deepens. These moths hover on rapidly beating wings and are sometimes mistaken for hummingbirds. In fact one of them is named the Hummingbird Sphinx.

The golden-orange of the black-eyed Susans are fading from the scene but the amethyst blossoms of New England asters with their rich yellow centers are beginning to bring their loveliness to the countryside.

Goldfinches and goldenrod seem to complement each other. As the goldenrod takes its place in the scheme of things, the finches are busily engaged in the task of preparing their fledglings to face the world on their own. Everywhere it seems, can be heard the happy and varied songs and notes of the goldfinches. They sometimes appear just as cheerful attending to their parental duties as when they were enjoying their vacation earlier in the season.

Catbirds mew from the thickets and swallows gather on the utility wires, twittering and chattering as though enjoying one last holiday before starting their journey southward.

A beautiful blue haze of chicory lines the edges of country roads, proclaiming to the world that although Summer is far from over yet that Autumn is only biding its time to step out onto the stage of the seasons.,

Warm evenings, a man often now notices baby toads hopping about beneath the porch light waiting for unwary insects to drop down within reach of their sticky tongues.

While the toads work on the ground several varieties of leathery-winged bats of varying sizes patrol the evening sky, harvesting the abundant crop of noxious insects that this time of year provides,

The last days of August blend into the first days of September with no apparent difference between them. Still, a countryman can sense a subtle change in all the natural world. That great Wheel of The Seasons is turning from maturity towards that of harvest. So it is that those promises of Spring are rapidly approaching their complete fulfillment.

Summer is on the wane and a relatively dry summer has created, in a sense, an early autumn. The elderberries are shriveling up, what haven't been consumed yet by the birds, and grapes are ripening a mite early this year.

Goldenrod and asters brighten the fields and woods-edges but already many of them have blooms that have turned dull in color as the seeds mature. Silvery yarrow blossoms have turned a lead-gray and the wild carrots have dried and curled into their characteristic "birds-nest" shape.

Sumacs toss colorful banners into the afternoon breeze and Virginia creeper vines glow scarlet as the westering sun shines through them as they climb into saplings and a nearby spruce tree,

Teasels and burdocks are brown and spiny, their blooms now replaced by maturing seeds. Milkweed pods swell more and more as the summer draws to a close, the maturing seeds within crowding one another as they grow.

Bluejays call raucously to one another as they forage about in the late summer sun among the oaks and the pines and in the hedgerows.

Chipmunks chuck sonorously from stonepiles or atop old stumps in the woodlots. They often may be seen scampering about, their cheek pouches bulging with various items of food to be stored away in their underground pantries.

Autumn is steadily drawing nearer each day by the calendar, but a countryman already notices changes in the atmosphere, the subtle changes in the shifting winds and the cloud formations, There is a chill in the night air now that was lacking a couple weeks ago and the mornings are noticeably cooler with a gentle hint towards first frost in the low-lying valleys.

Gone from the barnyard and the utility wires are the familiar twitterings of the barn swallows, and gone too are many of the songbirds which winter in warmer climes.

As the sun sets and cold shadows creep over the land the fiddling of the crickets becomes increasingly slower in cadence. The evening star shines brightly above the eastern ridges and later on a quarter moon will shine silver in a star-studded sky.

# AUTUMN

Another Autumn has arrived once more. The sun rises later each morning and sets earlier each evening. For a brief time the length of day and night will be equal just as it was back in March. All too soon the days will become increasingly shorter and shadows will gather across the countryside earlier each evening.

For now, purple asters add their vibrant colors to fading fields of goldenrod. A countryman even found a few clusters of butter-and-eggs blooming bravely here and there.

It was very pleasing to a countryman to see a small number of monarch butterflies fluttering and floating their way in a southerly direction as the balmy late summer days merge into the cooler days of Autumn.

During the warmer days and evenings, katydids and other summer insects join together to provide the superb orchestration noted each year as summer draws gently to a close.

The little tree frogs so well known as “Spring Peepers” are now chirping their way through the woodlots as yet another indicator of approaching Autumn. Within a few more weeks they will have arrived at their wetland winter quarters.

It seems to a countryman that each succeeding year the seasons roll around much more quickly than they did a half century and more ago. No doubt that is yet another sign of his advancing years. Be that as it may, each season brings its own particular pleasures, each dawning day its own delight and each setting sun its own serenity. Man may modify the land, and with his pollution to some extent the seasons, yet they continue to move, unhurriedly and reliably, across the landscape of the year.

Most evenings of late, small flocks of a dozen or so geese honk their way to resting places for the night. Their increased movements and clarion calls assure a countryman of Summer’s passing.

With the passing of the geese, so does the day and shadows shroud the land with their ever-increasing darkness.

The insect orchestra has commenced their evening program and a small toad waits patiently beneath an outside light for its evening meal to arrive.

When Autumn arrived it closed the door to Summer with a bang. It was as though Autumn had said, "I have arrived and now I am in charge". Even the geese seem to have taken those words to heart. Each day, it seems, a flock or two can be seen or heard heading south, their wild cries floating down from the heights above.

Meanwhile, the local geese are moving about in between the lakes. Perhaps they are anxious for the cornfields to be harvested so that they can glean the left-overs and put on much needed fat for the winter.

A pair of killdeers is spending time lately about the margins of the farm pond. A short distance away, two flickers are foraging on the ground and they call "wick-up!" several times as a man passes by.

The calls of the bluejays seem to echo each other across the countryside and their voices add much to the autumnal atmosphere.

A robin perches on a fence post in the morning sun. A countryman notices that, although its breast is brick red, yet there are a few residual spots between the throat and the upper portion of the breast. These were juvenile spots, not to be confused with the normal dark throat stripings of an adult. This youngster is maturing just in time for the cold season just ahead in another month or so.

Goldfinches gather in the maturing weed patches. A catbird mews from a hedgerow. While watching a cardinal, a man notices an immature magnolia warbler flitting about the branches of a hickory tree. Warblers eat and rest during the day and migrate southward at night. Other songbirds follow this pattern as well. Still others work their way southward during the day and rest at night.

A countryman was fortunate to see several black swallowtail butterflies along with a few monarchs and two viceroy butterflies. The latter are a bit smaller than the monarchs and mimic them in color and pattern, thus diminishing the chances of being eaten.

Several chickadees come within a foot or two of a man's face and by their actions and voices, they act as though they expected a hand-out. When none was forthcoming, they moved slowly away, their cheery calls adding to the pleasantness of the day.

When evening skies are clear, the stars shine brilliantly, like diamonds across the dark expanse. The Big Dipper lies low above the northern horizon at this time of the year.

Soft notes from migrating songbirds fall gently upon the ears of a countryman as he gazes meditatively into the star-studded autumn sky.

First frosts and katydids are an unlikely combination, nevertheless several of the little fiddlers were making merry on several mild autumn evenings since Jack Frost scattered his cold crystals across the countryside.

Members of the insect orchestra that have found secluded shelter from the penetrating cold may well survive several weeks into early autumn. A countryman listens carefully for one of the little fiddlers until it becomes so cold that they have expired or hidden themselves away into the shelter of a warm building.

A man remembers a cricket in the kitchen wood box on a frosty, late fall evening. The crackling of the fire in the kitchen stove, the faint smell of wood smoke and the cheerful chirping of the cricket from the wood box, all are memories from the distant past that come easily to a man's mind.

October is the month of trees arrayed in splashes of orange, red and yellow and all the many shades in between. October is the month of orange pumpkins lying in the fields, waiting to become Jack O' Lanterns, mouth watering pies and, on rare occasions, pumpkin pudding which is nothing more than pie filling without the benefit of pie tin or crust. A countryman's mouth will water at the thought of spicy pies and puddings.

Falling leaves and migrating geese are also a major part of October. Shuffling through spicy-scented and colorful leaves is an activity that a man relishes each autumn as well as listening for the honkings of migrating geese as they pass high overhead in compact vees or strung out in long, wavering skeins.

Mellow days, heralded by frosty mornings, usher in the tenth month. The promises of the Spring, followed by the developments of Summer, are all realized in the fulfillings of Autumn as it reveals its bounties in preparation for the cold months that lie ahead, when the land can finally claim its own well-earned rest.

The countryside becomes more colorful as October reaches its midpoint. Rich shades of orange, red and yellow work their way into the green fabric of the long ridges between the lakes. The flaming foliage of the sumacs brightens the hedgerows and road-edges. Virginia creeper burns its way up tree trunks and outbuilding walls in the afternoon sun.

Chipmunks chip and sometimes chuck from the recesses of woodlots or the seclusion of old stone walls. Gray squirrels argue emphatically about individual rights to certain nut trees while bluejays scream harshly that everyone is a thief and at the same time waiting and watching so as to steal from their neighbors.

Robins have re-appeared in small flocks and they communicate cheerily to one another and to everyone who will take the time to listen.

Bluebirds, absent for several weeks, fly about the countryside in pairs or in small family groups Their soft, sweet warbles are always music to a countryman's ears.

A walk through a brushlot often proves rewarding to a man. At this time of the year white-throated sparrows chirp contentedly from brushy areas and even a towhee voices its opinion from somewhere nearby.

Towhees or chewinks, as they are often know, are more frequently heard than seen. They prefer to forage on the ground, protected by dense undergrowth. Sometimes their vigorous scratchings can be heard. In years past a countryman has watched them digging, using both feet at once in a most efficient manner, making the leaves fly in every direction.

Walking past a wild apple tree, a man sees a quick movement in the lower branches. Upon closer observation with binoculars, a female ruby-crowned kinglet is seen inspecting branchlets and the undersides of the leaves for insects or their eggs or perhaps larvae.

A countryman looks forward each year to the warm, sun-filled October days with their colorful countrysides and bright blue skies. He welcomes the sights and sounds of migrating birds and other wild creatures preparing for winter as well as those which will remain in the area and active during the cold months ahead.

A man relishes the bright colors of purple asters and goldenrods that have escaped Jack Frost's frigid fingers. These usually are the late bloomers in sheltered, northern exposures or are protected by surrounding trees and shrubs.

October, a countryman thinks, is one of the finest months of the entire year, a month to be carefully savored even as each day quickly slips away on brightly colored feet.

Fog hugs the distant ridges with its damp embrace and a light mist falls gently, caressing the bosom of the earth with its tender touch. Everything is soaked; beads of water hang from every branch and leaf. The world is wrapped in silence much as the ridges are enrobed with the fog. Two jays scream almost in unison and the silence is broken. A dog barks in the distance and a cow bawls from a nearby pasture. Then all is once more quiet.

Fields of goldenrod are no longer bright with color. Their once kingly crowns bow before the mist, sodden and brown yet ultimately triumphant, for where once golden blossoms bloomed now lie treasures of a different sort. This treasure is the seeds of a future generation, an assurance that there will be fields of goldenrod in the year to come.

New England asters lift their purple heads to the mist as if in gratitude for the refreshing moisture. Their royal colors brighten the otherwise drab fields of now mature flowers, their spirits seemingly uplifted rather than dampened by the gently falling mist.

Here and there in the woodlots soft maples splash their crimson colors against a mostly-green background. Staghorn sumacs flout their flaming banners to the fog-shrouded countryside.

A countryman walks towards home, breathing deeply of the cool, moist air and savoring its freshness and unadulterated purity. There is much to be said for sun-filled, warm October days but a man can also sense the richness of color and softness of mood of a fog-enshrouded and mist-laden autumn day.

Autumn winds shake the trees and multi-hued leaves scatter to the bosom of the earth, covering it with a colorful blanket. Black walnuts lose their grip on their natal branches and briefly shower earthward, hitting the ground with resounding thumps.

Hawks, migrating southward, sail with the wind on favorable days when winds blow out of the north. Local red-tailed hawks use the winds to their advantage while hunting for something to fill their empty stomachs. They often also may be seen perched in trees or atop utility poles along roadways as they scan the ground for prey.

As a countryman watches several robins and listens to their conversational chirpings, he hears the dulcet voices of bluebirds. He looks in vain for them but they remain hidden from view. Only their soft warbles betray their presence and this is enough for a man as he listens to them and their robin cousins perched nearby.

During a late, warm afternoon a katydid rasps out its solo notes. A cricket chirps in an emphatic manner from a vineyard row as a man walks by.

Flickers call “clee-up!” back and forth. One lands in the top of a tree and teeters precariously on its perch as the wind flows through the branches.

A downy woodpecker goes on an inspection tour of the sumacs in the hedgerow, hitching its way up first one crooked trunk and then another. A white-breasted nuthatch calls nasally as it walks head-first down the rough bark of an aging maple.

Later in the afternoon, as a man works on a project on his porch, he is surprised to hear, almost at his elbow, the sharp soprano voice of a peeper. The tiny tree frog seems to be hidden in one of the evergreen shrubs growing close to the porch. As a man continues with his work, the strident calls continue, only interrupted by brief periods of silence.

Two white-crowned sparrows forage about the shrubbery as a man finishes up his project. Eventually they move elsewhere.

A man picks up his tools and materials and leaves the porch area to the little tree frog who voices its farewell in the same vein as it first announced its presence.

Autumn lengthens into the eleventh month and its previously mild and easy-going nature now takes on a more severe attitude, using a sharp north wind as emphasis. It is as if Autumn is advising the land and its inhabitants that harder times lie just around the corner.

At this time of the year the annual autumn rains can be expected and indeed they are much needed before the ground freezes. When the rains come in November they often change suddenly into snow pellets that rattle among the leaves on the forest floor. On the other hand, showers may quickly convert to frozen flakes of snow that settle swiftly earthward.

The vegetation has now been seared by the frosts and freezes. Weed patches are brown and sere and play host to goldfinches, juncos and sparrows.

Local geese honk their way back and forth between the lakes and whatever fields of corn that have been at least partially harvested.

Crows harass a hawk in a wooded area and make it very difficult for the raptor to concentrate on its visual search for food.

Three deer feed at the edge of an alfalfa field for some time until, for some reason, they decide to move back into the seclusion of the nearby woodlot.

The raw air of the day makes a countryman decide to head towards the comforts of home. A song sparrow calls from an overgrown brush pile and a gray squirrel, camouflaged against the gray trunk of an oak tree, hopes in vain that the passing man will not see it. It remains motionless, only its tail, fluttering in the breeze, betrays its presence. A bit further on, a little red squirrel clings sideways to the curling, gray bark of a hickory and watches as a man approaches. Suddenly, it races up the tree and out of sight.

A countryman turns up his collar against the bite of the wind and continues on his way, enjoying the sound and scent of the dried leaves that cackle beneath his feet.

Made in United States
Cleveland, OH
05 November 2026

47664321R00183